SpringerBriefs in Genetics

Marién Pascual · Sergio Roa

Epigenetic Approaches to Allergy Research

 Springer

Marién Pascual
Oncology Division
Center for Applied Medical
 Research (CIMA)
University of Navarra
Pamplona
Spain

Sergio Roa
Oncology Division
Center for Applied Medical
 Research (CIMA)
University of Navarra
Pamplona
Spain

ISSN 2191-5563   ISSN 2191-5571   (electronic)
ISBN 978-1-4614-6365-8   ISBN 978-1-4614-6366-5   (eBook)
DOI 10.1007/978-1-4614-6366-5
Springer New York Heidelberg Dordrecht London

Library of Congress Control Number: 2012956299

Printed on acid-free paper

Springer is part of Springer Science+Business Media (www.springer.com)

Preface

The standard dogma back in the twentieth century was that the human genome was 99 % "junk DNA", and that most of the non-coding sequences across the genome did not hold much interest. However, many scientists across the world relentlessly confronted such plain hypothesis. As an old friend and colleague at the University of Salamanca used to say, "if it exists, it can't be without a meaning". As a recent outcome from such dissatisfaction accepting the existence of boring "junk DNA", a worldwide consortium of scientists just published in September of 2012 the first results of the ENCODE project [1], an encyclopedia of DNA elements, including protein, RNA, and regulatory elements that have a functional impact in the circumstances in which genes are active. This roadmap of the genome can be explored by readers through the *Nature* ENCODE portal (http://www.nature.com/encode/#/threads), and for many it has been considered, among other things, a tremendously interesting repository of epigenetic information. We wrote this *Briefs* with the idea in mind that this is the time for clinical researchers and allergologists to connect with the ENCODE goals and use them as a first GPS to better understand asthma and allergy.

Therefore, this *Springer Briefs* is intended as an introductory text for a broad spectrum of students, clinicians, and scientists who want to get a quick overall understanding of the increasing evidences of crosstalk between the fields of allergy and epigenetics. The most relevant immunological aspects of allergy and its pathophysiology are treated early in this book. Then, the variety of epigenetic elements (i.e. DNA, RNA and protein) that may have a critical impact in gene regulation are discussed, including a concise description of the major current technologies to study them. Finally, we review several hypotheses and recently published data, which support the increasing interest on the profound impact that environment and epigenetics have in the pathogenesis of complex diseases like asthma and allergy.

The fruitful collision of three perspectives is the origin of this book. On one side, Marién Pascual, Ph.D, conducted the *epigenetic perspective*. She got interested in epigenetics and allergic diseases early in her career at the Allergy Department of the University Hospital of Salamanca (University of Salamanca, Spain). This interest was further cultivated when she joined the group of

Dr. John Greally (Albert Einstein College of Medicine, New York, USA), a lab devoted to the study of DNA methylation. An intense collaboration between both groups culminated in several papers about epigenetics and allergy. Then, here comes the second perspective, the *editorial perspective* that made Meredith Clinton, Assistant Editor at Springer, believe that a volume on the role of epigenomics in asthma and allergy would be a fantastic addition to the Springer book program and the community at large. Finally, Sergio Roa, Ph.D, got onboard this project to bring his *immunological perspective* to the project. Such collaboration responded to many years as colleagues at the Albert Einstein College of Medicine in New York and numerous discussions about the role of DNA methylation in B cell biology. Inspired by Marién's passion for epigenetics and new technologies, and encouraged by Dr. Sandeep N. Wontakal and Dr. Richard Chahwan from the Einstein Relativity Club, Sergio has tried to bring into this monograph his AID-biased perspective and illustrations of the immune response.

We would like to give great thanks to all those people in Salamanca (especially to Dr. María Isidoro-García, Dr. Ignacio Dávila, and Dr. Félix Lorente), in New York (especially to Dr. Masako Suzuki, Dr. John Greally, and Dr. Matthew D. Scharff) and Pamplona (especially to Dr. Felipe Prósper and Dr. José Ángel Martínez-Climent), as well as to our family and friends, who supported and advised us on how to make this project better. We want also to give thanks to Dr. Esteban Pascual Pablo, MD, who carefully reviewed the manuscript.

Our deepest hope is that the ideas that we discuss in this book have the chance to inspire and fuel the interest of the reader to pursue new ways of exploring epigenetics and allergy, as another step, even if it is small, toward a better understanding and treatment of allergic disease.

Marién Pascual
Sergio Roa

Reference

1. Bernstein BE, Birney E, Dunham I, Green ED, Gunter C, Snyder M. An integrated encyclopedia of DNA elements in the human genome. Nature. 2012;489(7414):57–74. doi:10.1038/nature11247.

Contents

Chapter 1
Introduction

All living organisms are continuously exposed to the deleterious risk of being "attacked" by both pathogenic substances and toxins, or other organisms such as bacteria, fungi, viruses, and parasitic protozoans. As if this was not enough, sometimes usually harmless substances known as allergens are able to stimulate the immune system, as we review in this issue, promoting hypersensitivity allergic reactions.

To protect themselves against this plethora of pathogens, multicellular organisms have developed a first line of protection known as *innate immunity*. This kind of immunity is mediated by both cellular and humoral mechanisms, and triggers inflammation to response against the pathogenic assault. Despite of its specificity, however, it is characterized by the absence of memory, even if the antigen has been previously encountered. In addition to this line of defense, vertebrates have evolved an additional protective system characterized by its immunological memory and known as *adaptive immunity*. This system is highly specific and has plasticity because of the complex affinity diversification and maturation processes experienced by the receptors at the surface of the immune cells. As a consequence, subsequent encounters with the same antigen that has been previously presented to the adaptive immune system, will elicit a quicker and stronger immune response. A tight equilibrium of these immune responses can be broken in some instances, which happens when the activation threshold is too low (hypersensitivity) or when self-molecules can erroneously activate the systems (autoimmunity).

The twentieth century has provided a deep understanding of the cellular components and molecular mechanisms that characterize both the innate and adaptive immune responses. Indeed, we understand now better how antibodies and cellular receptors are constructed. However, much is still to be discovered in this twenty-first century about the fine regulation, communication, and function of the immune system. As an example, the discovery of the mechanisms involving the Toll (in 1996) and Toll-like receptors (in 1998), which trigger the activation of the innate immunity, as well as the discovery of the dendritic cells (in 1973), which have a unique capacity to mediate the communication between innate and adaptive immunity, have just been recently awarded the 2011 Nobel Prize in Physiology or Medicine. These discoveries have triggered an explosion of research from fundamental to translational biomedical areas. The authors of this *Springer Briefs*

M. Pascual and S. Roa, *Epigenetic Approaches to Allergy Research*,
SpringerBriefs in Genetics, DOI: 10.1007/978-1-4614-6366-5_1,

believe that epigenomics is another field rapidly expanding with the advent of current amazing technological developments. It is our hope that this brief review will inspire and help the reader to better understand that epigenetic mechanisms are critically involved in allergic responses, and deserve to be carefully explored as to make possible in the future the development of new methods for preventing and treating allergic diseases.

Chapter 2
Immune System and Atopic Disorders

2.1 Asthma and Allergy

The terms "complex" or "multifactorial" are used interchangeably to refer to diseases that are obviously not the result of a single mutation or an environmental aggression. The genetic load of these diseases is unquestionable, and numerous studies have demonstrated both its heritability and the influence of certain anti-inflammatory factors, but both have proved to be insufficient to the complete understanding of their prevalence and patterns of heritability.

Allergic rhinitis, atopic dermatitis, allergic asthma, food allergy, anaphylaxis, or contact dermatitis are examples of complex or multifactorial diseases for which much is yet to be understood. The term "allergy" was first used by Clemens von Pirquet in 1906 to define the unusual tendency of some people to develop reactivity symptoms or "hypersensitivity reactions" when exposed to seemingly innocuous substances. Atopic diseases, from the Greek *atopos* meaning "out of place", are associated with the production of specific IgE antibodies and with the expansion of specific T cell populations, which are reactive against what normally would be harmless substances.

The prevalence of allergic diseases has increased around 75 % during the last three decades [1]. This rise in the epidemiological trend has not found a satisfactory explanation yet, although there is a general consensus that is not due solely to a genetic explanation. This observation has been subject of intense speculation, and the role of certain environmental factors has been studied. To this intriguing scenario we have to add that the pathogenesis of the disease as well as the contribution of genetic factors is still poorly understood [2].

Today it is thought that about 300 million people worldwide suffer from asthma [3, 4]. According to the American Academy of Allergy, Asthma and Immunology (www.aaaai.org), it is thought that more than half (54.6 %) of Americans are positive for one or more allergens [5, 6], and that about 50 millions suffer from some kind of allergy, including allergies to food, drug, latex, insects, pollen, mites, or skin and eye allergic diseases. Allergic diseases in the US occupy the fifth position among the most common chronic diseases, and the third position among the most common chronic diseases in children under 18 years [7]. In a pioneer study that was conducted in Britain in 1992, the comparison of children from 1964 to

M. Pascual and S. Roa, *Epigenetic Approaches to Allergy Research*,
SpringerBriefs in Genetics, DOI: 10.1007/978-1-4614-6366-5_2,
© The Author(s) 2013

those from 1989 showed that asthma prevalence had increased from 4.1 to 10.2 %, respectively, whereas rhinoconjunctivitis to pollen raised from 3.2 to 11.9 %, and atopic dermatitis from 5.3 to 12 % [8]. In European countries like Spain, it has been estimated that the prevalence of allergic diseases in adults is around 21.6 % (about one in five individuals will experience some kind of allergy-related disease during their life time) [9]. The prevalence is higher in women than in men and the number of cases is greater in larger cities. Among the main causes inducing allergy are pollens, domestic dust mites, and drugs. Allergic diseases are one of the most frequent reasons for pediatric medical visits, where children exhibit symptoms that can limit their daily activities in more than 40 % of the cases.

2.2 Classification of the Allergic Reactions

According to the traditional classification proposed by Gell and Coombs in 1963, hypersensitivity reactions can be divided into four groups [10] (Table 2.1). Although our current knowledge of the immune system has widely increased, this classification is still in use, at least in teaching terms. In the clinical practice, however, its use has been challenged since it does not include all possible allergic reactions caused by medical drugs, and sometimes both humoral and cellular responses can occur at the same time [11, 12]. Other scientists have proposed the existence of additional types of hypersensitivity [13, 14]. However, for the clarity and simplicity of our exposition of the fundamental concepts that inspire the classification of hypersensitivity reactions, we will start here by discussing the traditional one, to give a quick overview of the different predominant scenarios (see Table 2.1), and then we will annotate some of the improvements and subtypes.

In a more precise way, anaphylactic **Type I** reactions or immediate reactions those that refer to an IgE-mediated response [15]. These reactions take place in seconds to minutes after contact with the allergen, due to the release of preformed

Table 2.1 Original Gell and Coomb's classification of hypersensitivity reactions

Antigen-mediated hypersensitivity			Cell-mediated hypersensitivity
Type I	Type II	Type III	Type IV
IgE-mediated release of histamine and other mediators from mast cells and basophils	Involve IgG or IgM antibodies bound to cell surface antigens and complement fixation	Involve circulating antigen/antibody immune complexes and complement fixation	Mediated by T cells rather than by antibodies
↓	↓	↓	↓
Immediate hypersensitivity reactions, allergic rhinitis, bronchial asthma	Cytotoxic hypersensitivity reactions, thrombocytopenia	Immune-complex reactions, serum sickness	Delayed hypersensitivity reactions, contact dermatitis

mediators from mast cells and basophils. Most allergens that can trigger this response have low molecular mass (ranging between 6 and 120 kilodaltons, kDa) [16] and are highly hydrophilic, capable of penetrating the human body through mucosal areas in the respiratory and digestive tracts [11]. In sensitized people, even trace amounts of allergen are capable of triggering an allergic response.

Antibodies of the IgG or IgM isotypes mediate **Type II** or cytotoxic reactions. These antibodies are able to recognize antigens that are located on cellular surfaces. Type II reactions involve the formation of immune complexes and the interaction of these complexes with the complement system, Fc-IgG receptors present on macrophages and Natural Killer cells, among others. The affected cells are removed in minutes. Among the diseases associated with complement activation and type II hypersensitivity responses are drug-induced cytopenias [12]. The new improvements in the original classification from Gell and Coomb, now consider that **Type IIa** corresponds to the former Type II, where the inflammatory response results in cell death mediated by activation of the complement, phagocytosis, or cytolysis following the binding of antibodies to cell surface antigens on the target cell. On the other hand, the newly defined **Type IIb** is mechanistically distinct and refers to the inflammatory response resulting in cell death which is mediated by the direct binding of antibodies to cellular receptors [14].

Type III reactions are mediated by antigen–antibody immune complexes that are insoluble in the bloodstream. Despite being itself a physiological process, if these complexes are not properly removed by phagocytosis in the spleen or other lymphoid organs, they may be deposited on the walls of the circulatory system, skin, or joints. Remaining immune complexes are able to trigger the activation of the complement system and the recruitment of other immune cells.

In the pathophysiology of hypersensitivity **Type IV** reactions or delayed reactions, T cells react independently of antibodies against self or foreign antigens associated with tissues or cells. This type of reaction has been later subdivided into four subtypes according to the type of T cells involved [17, 18]:

- **Type IVa** hypersensitivity reactions are mediated by CD4$^+$ Th1 T lymphocytes, which activate macrophages through the secretion of INF-γ (probably including also other cytokines such as TNF or IL-18), promoting the B cell-mediated production of antibodies related to the complement system (IgG1, IgG3), stimulating the inflammatory response, and activating CD8 $^+$ mediated T cell responses [17, 18].
- **Type IVb** hypersensitivity reactions are mediated by CD4$^+$ Th2 T lymphocytes, which secrete IL-4, IL-5, and IL-13. These cytokines promote the production of IgE and IgG4 by B cells and have the potential to activate eosinophils and mast cells [17, 18]. High levels of secreted IL-5 promote eosinophilic inflammatory responses characteristic of many adverse drug reactions [19].
- In **Type IVc** hypersensitivity reactions, the cytotoxic CD8$^+$ T cells act as effectors by secreting cytolytic proteins (e.g. perforin) and serine proteases (e.g. granzyme B), thereby inducing apoptosis of the targeted cells [20]. In general, hypersensitivity reactions mediated by CD8$^+$ cytotoxic T lymphocytes

are considered more dangerous, since all cells expressing MHC-I complexes may be potential targets, as opposed to those reactions involving CD4$^+$ helper T lymphocytes that regulate the targeting of fewer cells expressing MHC-II complexes.

- **Type IVd** hypersensitivity reactions are characterized by the coordination of T cell-mediated neutrophilic inflammation. The principal mediators of this type of response are CXCL-8 and GM-CSF [17, 18].

2.3 Asthma Phenotypes

The first problem faced by epidemiological studies on asthma is its definition, which combines medical history and physical examination. The Global Initiative for asthma (GINA, www.ginasthma.org) defines the disease as "a chronic inflammatory disorder of the airways in which many cells and cellular elements play a role. The chronic inflammation causes an associated increase in airway hyperresponsiveness that leads to recurrent episodes of wheezing, breathlessness, chest tightness, and coughing, particularly at night or in the early morning. These episodes are usually associated with widespread but variable airflow obstruction that is often reversible either spontaneously or with treatment" [21].

This disease is found in both sexes and all ethnic groups, with variations both in symptoms and severity [22]. It has been proposed to consider asthma as a set of overlapping syndromes rather than as a single disease. The clinical features of the disease allow us to distinguish two types of asthma: atopic and non-atopic. Apart from being pathologically indistinguishable, both types of asthma are characterized by reversible airflow obstruction, wheeze with exertion, diurnal variation of bronchial tone and eosinophilia in sputum and peripheral blood [23].

2.3.1 Atopic Asthma

The inflammatory response in atopic asthma is associated with sensitization to certain allergens and, in most cases, with an elevation of total serum IgE. As we will review later, the asthmatic response is orchestrated by a typical Th2 T cell response, which is able to regulate and activate a wide range of other cells (B cells, mast cells, or eosinophils, among others) or by direct interaction with structural cells of the lung.

The Th2 response is mainly associated with the secretion of a set of interleukins, including IL-4, IL-5, and IL-13. The secretion of these cytokines results in the activation of adhesion molecules and the production of other cytokines and chemokines, which can trigger a process of recruitment and subsequent cellular degranulation of eosinophils, mast cells, and basophils, as well as B cell activation and concomitant production of IgE [11].

Many patients initially present only one type of allergic disease such as atopic dermatitis, but will eventually develop others such as allergic rhinitis, allergic asthma, or food allergy. This feature of allergic diseases is known as "atopic march" [24–26]. Rhodes and colleagues [27] conducted a study in the UK where they assessed the disease outcome of 100 infants on the basis that at least one parent was atopic. This study, which began in 1976, continued for a total of 22 years. The prevalence of atopic dermatitis was 20 % after the first year, dropping to 5 % at the end of the study. Meanwhile, the prevalence of allergic rhinitis was increasing over time, from 3 to 15 %. The respiratory distress described by the parents arose from 5 to 40 % at the end of the study. Allergic sensitization to a battery of common allergens reached 36 % of the participants. This study concluded that adults with asthma could begin wheezing at any age, but tended to be sensitized early in life. This typical sequence of clinical manifestations, characterized by the development of atopic dermatitis at an early age and later development of other diseases has been observed in numerous studies [28, 29].

2.3.2 Non-Atopic Asthma

Non-atopic asthma is characterized by the absence of specific IgE to a particular antigen in peripheral blood or bronchial mucosa. The serum IgE levels are often found within the normal range. It usually develops during middle age; remission is rare and generally tends to be more severe than atopic asthma. In this sense, there is a higher incidence of hypersensitivity to non-steroidal anti-inflammatory drugs (NSAIDs) [23].

Non-atopic asthma pathogenesis is also characterized by the development of an eosinophilic response, where eosinophils have important roles as proinflammatory cells driving the bronchoconstrictive response through the release of leukotrienes, oxygen metabolites, and toxic proteins, among other factors. Although this topic is out of the scope of this chapter, understanding the molecular mechanisms involved in non-atopic asthma is an exciting field for which other interesting reviews have been recently published and that we recommend to the interested reader [30, 31].

2.4 Pathophysiology of Allergy and Asthma

2.4.1 Immunological Basis of Allergic Reactions

Although there are different allergic reactions, all the allergic inflammatory responses are the result of a complex interplay of chemical and molecular signals and different immune cells, including dendritic cells, mast cells, basophils, eosino-phils, and lymphocytes (further discussed below). These cells produce a huge vari-ety of cytokines, chemokines, and reactive oxygen species that target other cells like epithelial, vascular, or airway smooth muscle cells, which will also become an important source of inflammatory mediators and signals. Depending on the affected cellular compartment, the associated symptoms will vary from asthma to allergic rhinitis/rhinosinusitis or atopic dermatitis.

2.4.1.1 Presenting the Allergen: Antigen Presenting Cells (APC)

Half of the 2011 Nobel Prize in Physiology or Medicine was awarded to Ralph M. Steinman for his discovery of the dendritic cells and their role in activat-ing the adaptive immune response (www.nobelprize.org). These cells, which are derived from hematopoietic progenitors, can be found in most tissues and are spe-cialized in taking antigens by endocytosis, processing them into smaller peptides, and presenting them to other immune cells through the major histocompatibility complexes (MHC) [32] (Fig. 2.1). Such antigen-presenting cells (APCs) have an exceptional ability to activate T lymphocytes, and they have opened to scien-tists new fields of research through the development of new ways of vaccination and treatment of diseases by boosting the immune response [33]. During allergic responses (see later in Fig. 2.6), APCs contribute to the progression of the sensiti-zation against the allergen by presenting the allergenic complexes peptide/MHC to T cells and activating its cytokine production. Interestingly, recent studies on the maternal transmission of asthma risk suggest that dendritic cells and DNA meth-ylation changes within these antigen-presenting cells might play a role in the con-genital susceptibility to allergic disease [34–36].

Fig. 2.1 The antigen presenting cell and the processing of allergens

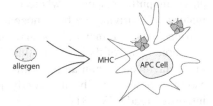

2.4.1.2 Mast Cells, Eosinophils, Basophils, and Neutrophils

Mast cell activation plays a pivotal role in initiating allergic reaction in the airways (Fig. 2.2), and their activation is responsible of the early phase of aller-gic reaction. It has been shown that monomeric IgE molecules have the ability to activate mast cells in several ways, from full mast cell activation (including degranulation) to enhanced mast cell survival [15, 37]. IgE can form clusters crosslinking FcεRI receptors and activate mast cell signaling, leading to a rapid

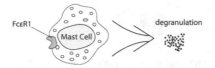

Fig. 2.2 Mast cell degranulation. Like mast cells, eosinophils, basophils, and neutrophils have shown to play a role during allergic responses mainly through the release of soluble mediators

release of inflammatory mediators (see later in Fig. 2.6). This activation will result in bronchoconstriction, vasodilatation, and plasma exudation, leading to the symptoms of asthma (wheezing and dyspnea) or allergic rhinitis. The role of mast cells during the development of late allergic responses remains still obscure [38].

Infiltration and accumulation of eosinophils in the bronchial mucosa are also responsible of many of the asthmatic symptoms. Eosinophilic activation depends on the release of IL-5 by Th2 cells [38], which leads to the release of cysteinyl leukotrienes. However, eosinophils are a less important source of inflammatory mediators when compared with mast cells. Interestingly, eosinophils are also able to act as APCs, producing several Th1 and Th2 cytokines [39].

Basophils are involved in the initiation as well as the maintenance of Th2 responses. Several studies show a role for basophils as APCs and it is known that they can release histamine, IL-4 and IL-13, or lipid mediators such as leukotrienes and prostaglandins, and express cysteinyl leukotriene mediators upon stimulation. Moreover, its presence in inflammated tissue during the allergic reaction clearly reveals a key role for basophils that has been underestimated [40]. Upon activation, basophils are able to skew Th2 responses toward allergens, secreting cytokines that support the development of IL-4-producing CD4+ T cells and of IgE-secreting B cells associated with the Th2 immune response [41].

Finally, the neutrophilic infiltration that is found in some patients with bronchial asthma might be driven by Th17 cells through the secretion of IL-17A, IL-17F, IL-21, IL-22, and IL-26, among other mediators [42, 43]. Although these observations support an important role of neutrophils in allergic reactions, their exact role is not completely understood yet and further studies are required.

2.4.1.3 T cells

In 1986, Mosmann and Coffman first described the existence of two different types of CD4+ populations, which are called T cell helper cell type 1 (Th1) or T cell helper cell type 2 (Th2) according to their pattern of cytokine secretion [44] (Fig. 2.3). The prevalent idea is that the immunological basis of atopic sensitization and allergic disease are the result of an inappropriate Th2 response to usually

Fig. 2.3 T cells (Th1 and Th2) and the secretion of cytokines. Regulatory T cells (Treg), both through cell-to-cell interactions and cytokine secretion, also play a critical role modulating the immune response and tolerance

harmless substances known as allergens. Some of the cytokines associated with this response are IL-4 and IL-13, important in the regulation of IgE production; IL-5, which contributes to the eosinophilic inflammation characteristic of allergic reactions; or IL-9 and IL-13, which are believed to be important for bronchial hyperresponsiveness [45] (see later in Fig. 2.6).

A new variety of T cells known as regulatory T cells (Treg) has recently entered in scene. This "new" cell type develops naturally in the thymus, through the stimulation of IL-2, TGFβ, and CD28 among others, and can also be induced in the periphery from naïve CD4 T cells (adaptive Treg) [46]. The best-characterized Tregs are CD4$^+$ lymphocytes that constitutively express high levels of surface receptor CD25 (the receptor for α chain of the IL-2: CD25hi T cells) [47]. Unlike CD4$^+$ lymphocytes, the lymphocytes CD25hi Treg do not proliferate or produce any cytokine after stimulation, but actively suppress the proliferation and cytokine production of other effector T cells. This "suppressor phenotype" is partly due to the expression of high levels of the transcription factor FOXP3 and expression of IL-10 and TGFβ [48, 49]. It is believed that these cells could play an important role in allergic diseases. Indeed, after separating the lymphocyte subsets CD4$^+$CD25$^+$ and CD4$^+$CD25$^-$ from peripheral blood of healthy individuals, it was found that in vitro the population CD4$^+$CD25$^-$ responded rapidly to antigenic stimulation, whereas in the presence of the CD4$^+$ CD25$^+$ population, the allergic response was inhibited [50].

2.4.1.4 B Cells

As mentioned before, humans, mice, and other vertebrate species are able to generate a highly diverse antibody response to protect themselves from infections and toxic substances in the environment. After B cells generate a large repertoire of antigen binding sites through combinatorial rearrangement of germline variable (V), diversity (D), and joining (J) elements [51] (see VDJ recombination later in Fig. 2.5), two processes called somatic hypermutation (SHM) and class switch recombination (CSR) further diversify the germline-encoded antibodies, which are often not protective, to produce high-affinity antibodies of all of the immunoglobulin isotypes. CSR to IgE is induced by the cytokines IL-4 or IL-13 secreted by T helper 2 (Th2) cells in humans [15] and CD40L or LPS plus IL-4 in mice [52]. In allergy and asthma, there is a significant bias toward the production of IgE by B cells and plasma cells in the respiratory tract mucosa [53, 54] (Fig. 2.4).

Fig. 2.4 The B cell and the production of IgE antibodies

Until recently, CSR had been thought to occur primarily in the germinal centers of lymphoid tissue, where the B cell expresses large amounts of the mutagenic enzyme activation induced deaminase (AID) [55, 56]. AID then induces mutation at the IgH and L chain V regions to initiate SHM, and at the switch regions (SRs) to initiate CSR. With the help of selection for B cells making higher affinity antibodies, these AID-induced mutations are responsible for affinity maturation and changes in specificity of the antibody response [57–63] and for the expression of different Ig isotypes that can carry out different effector functions throughout the body and in the mucosal spaces [64]. CSR to IgE and IgE synthesis, clonal selection and affinity maturation can also occur locally in the B cell population of local tissues and mucosas [65–68]. Both IgE and mast cells, which are concentrated in mucosal tissues, are thought to be a central part of that defense against pathogens [15]. Crosslinking of IgE and FcεRI, which is a high-affinity Fc receptor for IgE on mast cells and antigen-presenting cells (APCs), sensitizes these cells to allergens and promotes feedback mechanisms to produce more IgE by B cells [15]. These immune responses, however, may also result in immediate hypersensitivity, which characterizes allergic responses in different target organs such as the skin

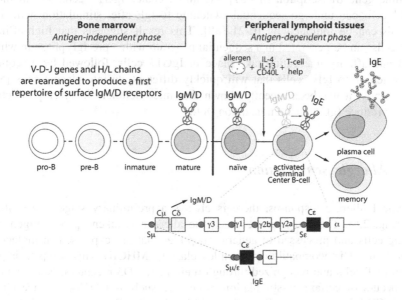

Fig. 2.5 B cell development and specific class-switch recombination processes to produce IgE antibodies

(atopic dermatitis or eczema), the nose (rhinitis), the lungs (asthma), and the gut (food allergic reactions) [37, 69].

Because AID is required for triggering switching to IgE production, it is important to understand not only how AID preferentially targets Ig genes to generate antibody diversity and why the rates of AID-induced mutation are higher in Ig genes than in most other genes [70], but also what regulates AID targeting specifically to the repetitive noncoding $S\mu$ and $S\epsilon$ switch regions (SR, see Fig. 2.5) during CSR to produce IgE antibodies. A number of not mutually exclusive mechanisms have been suggested to explain the targeting of AID [62, 71] including associated proteins, changes in chromatin structure to provide accessibility or specific recruitment through a histone code, and particular cis-acting sequences that bind transcription factors or other proteins that in turn recruit AID and its associated error-prone repair processes. Although some proteins that have been reported to interact physically with AID, including replication protein A (RPA), RNA polymerase II (RNA Pol II), MDM2, DNA PKcs, PKA, PTBP2, SPT5, the RNA exosome, or the splicosome-associated factor CTNNBL1 [72–82], all of these have very general functions and it is not immediately clear how they might be responsible for the preferential targeting of AID to V(D)J and SRs.

The molecular analysis of cis-acting regulatory elements in the human and murine IgE germline promoter region has identified several motifs that are bound by transcription factors upon IL-4 treatment. STAT6 binds one of these motifs [83, 84]. Recently, a cis-acting element with a PU.1 binding site that overlaps a NF-kB binding sequence was described and it has been suggested that the cooperation of either NF-kB or PU.1 with STAT6 mediates the IL4-induced activation of IgE germline gene transcription [85–87]. Germinal center IgG1$^+$ cells and memory IgG1$^+$ cells can undergo sequential switching to IgE after stimulation with IL-4 and this can be inhibited by IL-21 [88]. This result suggests that high-affinity IgE$^+$ cells can be generated in a sequential program with a pre-IgE phase in which SHM and affinity maturation take place in IgG1$^+$ cells, followed by a sequential switching to IgE$^+$ cells that will quickly differentiate into plasma cells [88]. Consistently, it has been recently shown that mice deficient in IgG1 production cannot produce IgE after immunization [89].

2.4.2 Allergic Sensitization

In Type I allergic responses, there is always a preliminary stage of sensitization (Fig. 2.6). Thus, in a first contact, the allergen is taken up by antigen-presenting cells and processed to generate peptides that are expressed in molecules of the Major Histocompatibility Complex class II (MHCII). This complex is presented to T cells and recognized through their TCR/CD3 receptors. Next, expression occurs of certain co-stimulation molecules such as CD154 (surface CD40 ligand, CD40L) located on the surface of T cells. CD40L molecules will bind to their corresponding receptor, CD40, present in the surface of B lymphocytes. This

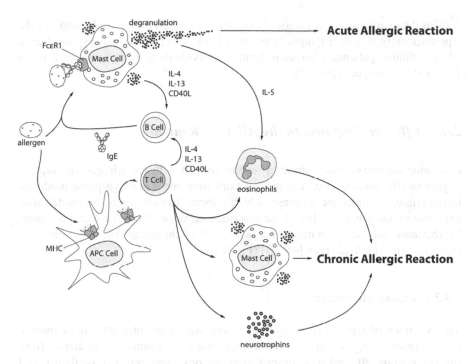

Fig. 2.6 Schema of the major cellular and signaling players during the allergic reaction. IgE production by B cells is activated by allergens and maintains mast cell and APC sensitization during allergic reaction

lymphocyte stimulation by TCR/CD3, MHCII/antigen, and CD40 and its ligand triggers a series of reactions that will culminate in the secretion of IgE by the B cell. Many B cell responses are governed by an integration of signals received by the B cell receptor (BCR) and other surface molecules. While in vivo studies have shown that the antigen recognition is necessary for a proper activation of B cells, the presence or absence of these other surface molecules determine the response of the cell after binding to the antigen through BCR [90]. One of the signaling cascades activated in B cells involve the induction of CD80/86, which will interact with CD28 on T lymphocytes and promote the expression of essential cytokines in the allergic response, such as IL-4, and IL-13 [15].

The binding of IL-4 and IL-13 to its corresponding receptor (IL-4R and IL-13R, respectively) present on the cell surface of B cells, activates the signaling cascade of STAT6 (Signal Transducer and Activator of Transcription 6). It is known that STAT6, in synergy with NF-kB, can activate the expression of AID, promoting class-switch recombination, and isotype switching from IgM to IgE [85, 87, 91, 92]. The IgE will then bind to the high sensitivity FcεRI receptors that are present in mast cells and basophils, triggering the release of inflammatory mediators responsible of the characteristic symptoms in allergic reactions. In patients with allergic diseases, the population of B lymphocytes and plasma cells in the respiratory tract mucosa is mostly committed to the production of IgE. Approximately, 4 % of B lymphocytes and 12–19 % of plasma cells express IgE in allergic rhinitis patients, whereas in healthy individuals these frequencies drop to 1 % or < 1 %, respectively [53].

2.4.3 Effector Response in the Allergic Reaction

Following the sensitization phase, a second contact with the allergen can rapidly trigger an effector response. Allergic reactions may occur in a two-phase mode: an initial phase or immediate response, which appears from seconds to minutes after exposure to antigen, and a late phase reaction that occurs between 4 and 8 h later. Furthermore, inflammation may occur as a result of chronic allergic reactions after repeated exposure to allergen [45, 93] (Fig. 2.6).

2.4.3.1 Immediate Reactions

The reactions of type I hypersensitivity occur within minutes after the exposure to a particular antigen, and involve the secretion of inflammatory mediators from mast cells in the affected sites. In sensitized persons, these cells expose their FcεRI surface receptors for high-affinity IgE. When several IgE molecules bind to antigens, FcεRI receptors aggregate and trigger a complex network of intracellular signaling which results in the release of mast cell granules. This process is known as degranulation and involves the merging of the cytoplasmic granules with the mast cell plasma membrane. These granules contain different types of biologically active products, including biogenic amines (e.g. histamine), proteoglycans (e.g. heparin), or serine proteases (e.g. triptasas or carboxypeptidases). Cytokines such as IL-4 and IL-13, or growth factors such as the tumor necrosis factor alpha (TNF-α) or Vascular Endothelial Growth Factor A (VEGFA) are also released as the degranulation progresses [94–97]. In addition, de novo synthesis of lipid mediators such as prostaglandins and leukotrienes also occurs in a matter of minutes. Arachidonic acid is metabolized by cyclooxygenases and lipoxygenases, and give rise to compounds known as prostaglandins such as prostaglandin D2 (PGD2), leukotrienes such as LTB4, and cysteinyl leukotrienes (cys-LTs) [94, 95].

The release of all these mediators is responsible for the characteristic symptoms of the immediate phase of allergic reactions: vasodilation (reflection of the activity of mediators acting in local nerves, causing erythema of the skin and conjunctiva), increased vascular permeability (resulting in the formation of edemas and tearing), bronchial smooth muscle contraction (and consequently airway obstruction and coughing), and increased mucus secretion (exacerbating the obstruction of the airways). In addition, these mediators may also stimulate nociceptors and nerve sensitivity of the nose [98], skin [99], or respiratory track [100], resulting in symptoms as characteristic as sneezing, itching, or coughing.

2.4.3.2 Late Stage

The allergic response stimulated by the IgE/allergen interaction in mast cells involves not only the immediate release of chemotactic factors, cytokines, chemokines, and growth factors previously synthesized, but also the synthesis of new inflammatory mediators that are released in a more gradually manner, constituting a delayed response [101]. During this late phase of allergic reactions, additional de novo synthesis of prostaglandins, leukotrienes, and various cytokines are induced in activated mast cells and basophils. Among the molecules that are released at this stage are TNFα, LTB4, IL-8 (also known as CXCL8) or CC2 chemokine ligand (CCL2). These factors are capable of recruiting other immune cells, activating cells responsible for innate immunity (through TNFα and IL-5), or activating various biological mechanisms that affect dendritic cells, T cells, and B cells through the activity of IL-10, TNFα, TGFβ or histamine [96, 102]. The clinical characteristics of these reactions highlight the contribution of both resident cells located at the damaged tissue and circulating cells that were recruited during the course of the inflammatory reaction. As an example, the associated vasodilation that occurs at late stages in atopic asthmatics by allergen-derived T cells peptides could be explained, at least in part, by the secretion of calcitonin and the recruitment of immunoreactive infiltrating inflammatory cells [103].

Late reactions usually occur 2–6 h after allergen exposure, with an elapsed peak around 6 or 9 h. The fact that such reactions do not appear in all sensitized individuals is not entirely clear, as it is not clear why in other patients there is not a clear distinction between the end of one phase and the beginning of the next [45].

2.4.3.3 Chronic Allergic Inflammation

When exposure to an allergen is produced in a continuous or repeated manner, many cells of the innate and adaptive immunity that are normally found in the bloodstream, may chronically infiltrate the affected tissues. The persistent inflammation that characterizes this type of exposures may cause structural changes in the tissues, compromising the correct functioning of the affected organs [104].

Although both the immediate and the delayed phase of allergic reactions can be studied relatively easily in human patients, most studies of chronic allergic inflammation have been performed in animal models. Therefore, it is yet not well understood how, after persistent exposure to an allergen, local tissue inflammation switches from an early or late phase to a chronic phase [102]. It is known that individuals with chronic asthma may have affected all layers of the respiratory tract, exhibiting changes in the epithelium and the number of goblet cells. This results in increased production of mucus, cytokines, and chemokines by epithelial cells, and ends up inducing tissue damage in the epithelium [105]. Respiratory viral infections with rhinovirus or influenza viruses are capable of producing a marked exacerbation of signs and symptoms of asthma [106], and might constitute an additional risk factor for subjects already at risk of developing allergies and asthma [107–109]. In cases of atopic dermatitis [110] and allergic rhinitis [111], as well as in asthma, chronic allergic inflammation is associated with tissue remodeling. In all cases, this remodeling may lead to persistent changes in the structural elements of the affected sites (such as increased vascularity) and substantially introduce alterations in the epithelial barrier function.

References

1. Devereux G. The increase in the prevalence of asthma and allergy: food for thought. Nat Rev Immunol. 2006;6(11):869–74. doi:10.1038/nri1958.
2. Vercelli D. Genetics, epigenetics, and the environment: switching, buffering, releasing. J Allergy Clin Immunol. 2004;113(3):381–386. doi:10.1016/j.jaci.2004.01.752.
3. Edgecombe K, Latter S, Peters S, Roberts G. Health experiences of adolescents with uncontrolled severe asthma. Arch Dis Child. 2010;95(12):985–91. doi:10.1136/adc.2009.171579.
4. Peters SP, Ferguson G, Deniz Y, Reisner C. Uncontrolled asthma: a review of the prevalence, disease burden and options for treatment. Respir Med. 2006;100(7):1139–51. doi:10.1016/j.rmed.2006.03.031.
5. Matasar MJ, Neugut AI. Epidemiology of anaphylaxis in the United States. Curr Allergy Asthma Rep. 2003;3(1):30–5.
6. Neugut AI, Ghatak AT, Miller RL. Anaphylaxis in the United States: an investigation into its epidemiology. Arch Intern Med. 2001;161(1):15–21.
7. Preparing a healthcare workforce for the 21st century: the challenge of chronic conditions. Chronic illness. 2005;1(2):99–100.
8. Ninan TK, Russell G. Respiratory symptoms and atopy in Aberdeen schoolchildren: evidence from two surveys 25 years apart. BMJ. 1992;304(6831):873–5.
9. Gaig P, Olona M, Lejarazu MD, Caballero MT, Dominguez FJ, Echechipia S. Epidemiology of urticaria in Spain. J Investig Allergol Clin Immunol. 2004;14(3):214–20.
10. Gell PGH, Coombs RRA. Clinical aspects of immunology. Oxford: Blackwell; 1963.
11. Averbeck M, Gebhardt C, Emmrich F, Treudler R, Simon JC. Immunologic principles of allergic disease. J Der Deutschen Dermatologischen Gesellschaft = J German Soc Dermatol : JDDG. 2007;5(11):1015–28. doi:10.1111/j.1610-0387.2007.06538.x.
12. Descotes J, Choquet-Kastylevsky G. Gell and Coombs's classification: is it still valid? Toxicol. 2001;158(1–2):43–9.
13. Rajan TV. The Gell-Coombs classification of hypersensitivity reactions: a re-interpretation. Trends Immunol. 2003;24(7):376–9.
14. Uzzaman A, Cho SH. Chapter 28: Classification of hypersensitivity reactions. Allergy and asthma proceedings : the official journal of regional and state allergy societies. 2012;33(Suppl 1):96–9. doi:10.2500/aap.2012.33.3561.
15. Gould HJ, Sutton BJ. IgE in allergy and asthma today. Nat Rev Immunol. 2008;8(3):205–17.
16. Wu CH, Lee MF. Molecular characteristics of cockroach allergens. Cell Mol Immunol. 2005;2(3):177–80.
17. Posadas SJ, Pichler WJ. Delayed drug hypersensitivity reactions - new concepts. Clin Exp Allergy. 2007;37(7):989–99. doi:10.1111/j.1365-2222.2007.02742.x.
18. Pichler WJ. Delayed drug hypersensitivity reactions. Ann Intern Med. 2003;139(8):683–93.
19. Hari Y, Urwyler A, Hurni M, Yawalkar N, Dahinden C, Wendland T, et al. Distinct serum cytokine levels in drug- and measles-induced exanthema. Int Arch Allergy Immunol. 1999;120(3):225–9.
20. Nassif A, Bensussan A, Dorothee G, Mami-Chouaib F, Bachot N, Bagot M, et al. Drug specific cytotoxic T-cells in the skin lesions of a patient with toxic epidermal necrolysis. J Invest Dermatol. 2002;118(4):728–33. doi:10.1046/j.1523-1747.2002.01622.x.

21. Bateman ED, Hurd SS, Barnes PJ, Bousquet J, Drazen JM, FitzGerald M, et al. Global strategy for asthma management and prevention: GINA executive summary. The European respiratory journal : official journal of the European Society for Clinical Respiratory Physiology. 2008;31(1):143–78. doi:10.1183/09031936.00138707.

22. Roy SR, McGinty EE, Hayes SC, Zhang L. Regional and racial disparities in asthma hospitalizations in Mississippi. J Allergy Clin Immunol. 2010;125(3):636–42. doi:10.1016/j.jaci.2009.11.046.

23. Kiley J, Smith R, Noel P. Asthma phenotypes. Curr Opin Pulm Med. 2007;13(1):19–23. doi: 10.1097/MCP.0b013e328011b84b.

24. Spergel JM. From atopic dermatitis to asthma: the atopic March. Annals of allergy, asthma and immunology : official publication of the American College of Allergy, Asthma and Immunology. 2010;105(2):99–106; quiz 7–9, 17. doi:10.1016/j.anai.2009.10.002.

25. Spergel JM. Epidemiology of atopic dermatitis and atopic March in children. Immunol Allergy Clin North Am. 2010;30(3):269–80. doi:10.1016/j.iac.2010.06.003.

26. Spergel JM, Paller AS. Atopic dermatitis and the atopic March. J Allergy Clin Immunol. 2003;112(6 Suppl):S118–27. doi:10.1016/j.jaci.2003.09.033.

27. Rhodes HL, Thomas P, Sporik R, Holgate ST, Cogswell JJ. A birth cohort study of subjects at risk of atopy: twenty-two-year follow-up of wheeze and atopic status. Am J Respir Crit Care Med. 2002;165(2):176–80.

28. Gustafsson D, Sjoberg O, Foucard T. Development of allergies and asthma in infants and young children with atopic dermatitis–a prospective follow-up to 7 years of age. Allergy. 2000;55(3):240–5.

29. Kulig M, Bergmann R, Klettke U, Wahn V, Tacke U, Wahn U. Natural course of sensitization to food and inhalant allergens during the first 6 years of life. J Allergy Clin Immunol. 1999;103(6):1173–9.

30. Kim HY, DeKruyff RH, Umetsu DT. The many paths to asthma: phenotype shaped by innate and adaptive immunity. Nat Immunol. 2010;11(7):577–84. doi:10.1038/ni.1892.

31. van den Berge M, Heijink HI, van Oosterhout AJ, Postma DS. The role of female sex hormones in the development and severity of allergic and non-allergic asthma. Clin Exp Allergy. 2009;39(10):1477–81. doi:10.1111/j.1365-2222.2009.03354.x.

32. Steinman RM. Decisions about dendritic cells: past, present, and future. Annu Rev Immunol. 2012;30:1–22. doi:10.1146/annurev-immunol-100311-102839.

33. Caminschi I, Shortman K. Boosting antibody responses by targeting antigens to dendritic cells. Trends Immunol. 2012;33(2):71–7. doi:10.1016/j.it.2011.10.007.

34. Lim RH, Kobzik L. Maternal transmission of asthma risk. Am J Reprod Immunol. 2009;61(1):1–10. doi:10.1111/j.1600-0897.2008.00671.x.

35. Fedulov AV, Kobzik L. Allergy risk is mediated by dendritic cells with congenital epigenetic changes. Am J Respir Cell Mol Biol. 2011;44(3):285–92. doi:10.1165/rcmb.2009-0400OC.

36. North ML, Ellis AK. The role of epigenetics in the developmental origins of allergic disease. Annals of allergy, asthma and immunology : official publication of the American College of Allergy, Asthma, and Immunology. 2011;106(5):355–61; quiz 62. doi:10.1016/j.anai.2011.02.008.

37. Gould HJ, Sutton BJ, Beavil AJ, Beavil RL, McCloskey N, Coker HA, et al. The biology of IGE and the basis of allergic disease. Annu Rev Immunol. 2003;21:579–628.

38. Barnes PJ. Pathophysiology of allergic inflammation. Immunol Rev. 2011;242(1):31–50. doi:10.1111/j.1600-065X.2011.01020.x.

39. Shamri R, Xenakis JJ, Spencer LA. Eosinophils in innate immunity: an evolving story. Cell Tissue Res. 2011;343(1):57–83. doi:10.1007/s00441-010-1049-6.

40. van Beek AA, Knol EF, de Vos P, Smelt MJ, Savelkoul HF, van Neerven RJ. Recent Developments in Basophil Research: Do Basophils Initiate and Perpetuate Type 2 T-Helper Cell Responses? Int Arch Allergy Immunol. 2012;160(1):7–17. doi:10.1159/000341633.

41. Chirumbolo S. State-of-the-art review about basophil research in immunology and allergy: is the time right to treat these cells with the respect they deserve? Blood transfusion = Trasfusione del sangue. 2012;10(2):148–164. doi:10.2450/2011.0020-11.

42. Alcorn JF, Crowe CR, Kolls JK. TH17 cells in asthma and COPD. Annu Rev Physiol. 2010;72:495–516. doi:10.1146/annurev-physiol-021909-135926.
43. Liang SC, Tan XY, Luxenberg DP, Karim R, Dunussi-Joannopoulos K, Collins M, et al. Interleukin (IL)-22 and IL-17 are coexpressed by Th17 cells and cooperatively enhance expression of antimicrobial peptides. J Exp Med. 2006;203(10):2271–9. doi:10.1084/jem.20061308.
44. Mosmann TR, Cherwinski H, Bond MW, Giedlin MA, Coffman RL. Two types of murine helper T cell clone. I. Definition according to profiles of lymphokine activities and secreted proteins. J Immunol. 1986;136(7):2348–57.
45. Kay AB. Allergy and allergic diseases. First of two parts. N Engl J Med. 2001;344(1):30–7. doi:10.1056/NEJM200101043440106.
46. Bluestone JA, Abbas AK. Natural versus adaptive regulatory T cells. Nat Rev Immunol. 2003;3(3):253–7. doi:10.1038/nri1032.
47. Sakaguchi S, Wing K, Miyara M. Regulatory T cells - a brief history and perspective. Eur J Immunol. 2007;37(Suppl 1):S116–23. doi:10.1002/eji.200737593.
48. Fontenot JD, Gavin MA, Rudensky AY. Foxp3 programs the development and function of CD4+CD25+ regulatory T cells. Nat Immunol. 2003;4(4):330–6. doi:10.1038/ni904.
49. Wan YY, Flavell RA. Regulatory T-cell functions are subverted and converted owing to attenuated Foxp3 expression. Nature. 2007;445(7129):766–70. doi:10.1038/nature05479.
50. Ling EM, Smith T, Nguyen XD, Pridgeon C, Dallman M, Arbery J, et al. Relation of CD4+CD25+ regulatory T-cell suppression of allergen-driven T-cell activation to atopic status and expression of allergic disease. Lancet. 2004;363(9409):608–15. doi:10.1016/S0140-6736(04)15592-X.
51. Maizels N. Immunoglobulin gene diversification. Annu Rev Genet. 2005;39:23–46.
52. Shen CH, Stavnezer J. Activation of the mouse Ig germline epsilon promoter by IL-4 is dependent on AP-1 transcription factors. J Immunol. 2001;166(1):411–23.
53. KleinJan A, Vinke JG, Severijnen LW, Fokkens WJ. Local production and detection of (specific) IgE in nasal B-cells and plasma cells of allergic rhinitis patients. The European respiratory journal : official journal of the European Society for Clinical Respiratory Physiology. 2000;15(3):491–7.
54. Dullaers M, De Bruyne R, Ramadani F, Gould HJ, Gevaert P, Lambrecht BN. The who, where, and when of IgE in allergic airway disease. J Allergy Clin Immunol. 2012;129(3):635–45. doi:10.1016/j.jaci.2011.10.029.
55. Muramatsu M, Kinoshita K, Fagarasan S, Yamada S, Shinkai Y, Honjo T. Class switch recombination and hypermutation require activation-induced cytidine deaminase (AID), a potential RNA editing enzyme. Cell. 2000;102:553–63.
56. Muramatsu M, Sankaranand VS, Anant S, Sugai M, Kinoshita K, Davidson NO, et al. Specific expression of activation-induced cytidine deaminase (AID), a novel member of the RNA-editing deaminase family in germinal center B cells. J Biol Chem. 1999;274(26):18470–6.
57. Di Noia JM, Neuberger MS. Molecular mechanisms of antibody somatic hypermutation. Annu Rev Biochem. 2007;76:1–22.
58. Teng G, Papavasiliou FN. Immunoglobulin somatic hypermutation. Annu Rev Genet. 2007;41:107–20.
59. Wang Y, Carter RH. CD19 regulates B cell maturation, proliferation, and positive selection in the FDC zone of murine splenic germinal centers. Immunology. 2005;22(6):749–61.
60. Longerich S, Basu U, Alt F, Storb U. AID in somatic hypermutation and class switch recombination. Curr Opin Immunol. 2006;18(2):164–74.
61. Casali P, Pal Z, Xu Z, Zan H. DNA repair in antibody somatic hypermutation. Trends Immunol. 2006;27(7):313–21.
62. Peled JU, Kuang FL, Iglesias-Ussel MD, Roa S, Kalis SL, Goodman MF, et al. The biochemistry of somatic hypermutation. Annu Rev Immunol. 2008;26:481–511.
63. Longo NS, Lipsky PE. Why do B cells mutate their immunoglobulin receptors? Trends Immunol. 2006;27(8):374–80.

64. Stavnezer J, Guikema JE, Schrader CE. Mechanism and regulation of class switch recombination. Annu Rev Immunol. 2008;26:261–92.
65. Snow RE, Djukanovic R, Stevenson FK. Analysis of immunoglobulin E VH transcripts in a bronchial biopsy of an asthmatic patient confirms bias towards VH5, and indicates local clonal expansion, somatic mutation and isotype switch events. Immunology. 1999;98(4):646–51.
66. Coker HA, Durham SR, Gould HJ. Local somatic hypermutation and class switch recombination in the nasal mucosa of allergic rhinitis patients. J Immunol. 2003;171(10):5602–10.
67. Gould HJ, Takhar P, Harries HE, Durham SR, Corrigan CJ. Germinal-centre reactions in allergic inflammation. Trends Immunol. 2006;27(10):446–52.
68. Takhar P, Corrigan CJ, Smurthwaite L, O'Connor BJ, Durham SR, Lee TH, et al. Class switch recombination to IgE in the bronchial mucosa of atopic and nonatopic patients with asthma. J Allergy Clin Immunol. 2007;119(1):213–8.
69. Kraft S, Kinet JP. New developments in FcepsilonRI regulation, function and inhibition. Nat Rev Immunol. 2007;7(5):365–78.
70. Liu M, Duke JL, Richter DJ, Vinuesa CG, Goodnow CC, Kleinstein SH, et al. Two levels of protection for the B cell genome during somatic hypermutation. Nature. 2008;451(7180):841–5.
71. Yang SY, Schatz DG. Targeting of AID-mediated sequence diversification by cis-acting determinants. Adv Immunol. 2007;94:109–25.
72. Basu U, Chaudhuri J, Alpert C, Dutt S, Ranganath S, Li G et al. The AID antibody diversification enzyme is regulated by protein kinase A phosphorylation. Nature. 2005;438(7067):508–11. doi:nature04255 [pii].
73. Nambu Y, Sugai M, Gonda H, Lee CG, Katakai T, Agata Y et al. Transcription-coupled events associating with immunoglobulin switch region chromatin. Science NY. 2003;302(5653):2137–2140.
74. Chaudhuri J, Tian M, Khuong C, Chua K, Pinaud E, Alt FW. Transcription-targeted DNA deamination by the AID antibody diversification enzyme. Nature. 2003;422(6933):726–30.
75. MacDuff DA, Neuberger MS, Harris RS. MDM2 can interact with the C-terminus of AID but it is inessential for antibody diversification in DT40 B cells. Mol Immunol. 2006;43(8):1099–108.
76. Wu X, Geraldes P, Platt JL, Cascalho M. The double-edged sword of activation-induced cytidine deaminase. J Immunol. 2005;174(2):934–41.
77. Conticello SG, Ganesh K, Xue K, Lu M, Rada C, Neuberger MS. Interaction between antibody-diversification enzyme AID and spliceosome-associated factor CTNNBL1. Mol Cell. 2008;31(4):474–84.
78. Stanlie A, Begum NA, Akiyama H, Honjo T. The DSIF subunits Spt4 and Spt5 have distinct roles at various phases of immunoglobulin class switch recombination. PLoS Genet. 2012;8(4):e1002675. doi:10.1371/journal.pgen.1002675.
79. Pavri R, Gazumyan A, Jankovic M, Di Virgilio M, Klein I, Ansarah-Sobrinho C et al. Activation-induced cytidine deaminase targets DNA at sites of RNA polymerase II stalling by interaction with Spt5. Cell. 2010;143(1):122–33. doi:S0092-8674(10)01065-2 [pii].
80. Vuong BQ, Chaudhuri J. Combinatorial mechanisms regulating AID-dependent DNA deamination: Interacting proteins and post-translational modifications. Semin Immunol. 2012;. doi:10.1016/j.smim.2012.05.006.
81. Nowak U, Matthews AJ, Zheng S, Chaudhuri J. The splicing regulator PTBP2 interacts with the cytidine deaminase AID and promotes binding of AID to switch-region DNA. Nat Immunol. 2011;12(2):160–6. doi:10.1038/ni.1977.
82. Basu U, Meng FL, Keim C, Grinstein V, Pefanis E, Eccleston J, et al. The RNA exosome targets the AID cytidine deaminase to both strands of transcribed duplex DNA substrates. Cell. 2011;144(3):353–63. doi:10.1016/j.cell.2011.01.001.
83. Fenghao X, Saxon A, Nguyen A, Ke Z, Diaz-Sanchez D, Nel A. Interleukin 4 activates a signal transducer and activator of transcription (Stat) protein which interacts with an interferon-gamma activation site-like sequence upstream of the I epsilon exon in a human B cell line. Evidence for the involvement of Janus kinase 3 and interleukin-4 Stat. J Clin Invest. 1995;96(2):907–14.

84. Hou J, Schindler U, Henzel WJ, Ho TC, Brasseur M, McKnight SL. An interleukin-4-induced transcription factor: IL-4 Stat. Science NY. 1994;265(5179):1701–7016.
85. Stutz AM, Woisetschlager M. Functional synergism of STAT6 with either NF-kappa B or PU.1 to mediate IL-4-induced activation of IgE germline gene transcription. J Immunol. 1999;163(8):4383–91.
86. Delphin S, Stavnezer J. Characterization of an interleukin 4 (IL-4) responsive region in the immunoglobulin heavy chain germline epsilon promoter: regulation by NF-IL-4, a C/EBP family member and NF-kappa B/p50. J Exp Med. 1995;181(1):181–92.
87. Messner B, Stutz AM, Albrecht B, Peiritsch S, Woisetschlager M. Cooperation of binding sites for STAT6 and NF kappa B/rel in the IL-4-induced up-regulation of the human IgE germline promoter. J Immunol. 1997;159(7):3330–7.
88. Erazo A, Kutchukhidze N, Leung M, Christ AP, Urban JF Jr, Curotto de Lafaille MA. Unique maturation program of the IgE response in vivo. Immunity. 2007;26(2):191–203.
89. Xiong H, Dolpady J, Wabl M, Curotto de Lafaille MA, Lafaille JJ. Sequential class switching is required for the generation of high affinity IgE antibodies. J Exp Med. 2012;. doi:10.1 084/jem.20111941.
90. Carter RH, Wang Y, Brooks S. Role of CD19 signal transduction in B cell biology. Immunol Res. 2002;26(1–3):45–54. doi:10.1385/IR:26:1-3:045.
91. Pate MB, Smith JK, Chi DS, Krishnaswamy G. Regulation and dysregulation of immunoglobulin E: a molecular and clinical perspective. Clin Mol Allergy. 2010;8:3. doi:10.1186/1476-7961-8-3.
92. Dedeoglu F, Horwitz B, Chaudhuri J, Alt FW, Geha RS. Induction of activation-induced cytidine deaminase gene expression by IL-4 and CD40 ligation is dependent on STAT6 and NFkappaB. Int Immunol. 2004;16(3):395–404.
93. Holgate ST, Polosa R. Treatment strategies for allergy and asthma. Nat Rev Immunol. 2008;8(3):218–30. doi:10.1038/nri2262.
94. Boyce JA. Mast cells and eicosanoid mediators: a system of reciprocal paracrine and autocrine regulation. Immunol Rev. 2007;217:168–85. doi:10.1111/j.1600-065X.2007.00512.x.
95. Boyce JA. Eicosanoid mediators of mast cells: receptors, regulation of synthesis, and pathobiologic implications. Chem Immunol Allergy. 2005;87:59–79. doi:10.1159/000087571.
96. Galli SJ, Kalesnikoff J, Grimbaldeston MA, Piliponsky AM, Williams CM, Tsai M. Mast cells as "tunable" effector and immunoregulatory cells: recent advances. Annu Rev Immunol. 2005;23:749–86. doi:10.1146/annurev.immunol.21.120601.141025.
97. Gilfillan AM, Tkaczyk C. Integrated signalling pathways for mast-cell activation. Nat Rev Immunol. 2006;6(3):218–30. doi:10.1038/nri1782.
98. Sarin S, Undem B, Sanico A, Togias A. The role of the nervous system in rhinitis. J Allergy Clin Immunol. 2006;118(5):999–1016. doi:10.1016/j.jaci.2006.09.013.
99. Cevikbas F, Steinhoff A, Homey B, Steinhoff M. Neuroimmune interactions in allergic skin diseases. Curr Opin Allergy Clin Immunol. 2007;7(5):365–73. doi:10.1097/ACI.0b013e3282a644d2.
100. Lalloo UG, Barnes PJ, Chung KF. Pathophysiology and clinical presentations of cough. J Allergy Clin Immunol. 1996;98(5 Pt 2):S91-6; discussion S6-7.
101. Rivera J, Gilfillan AM. Molecular regulation of mast cell activation. J allergy and Clin Immunol. 2006;117(6):1214–1225; quiz 26. doi:10.1016/j.jaci.2006.04.015.
102. Galli SJ, Tsai M, Piliponsky AM. The development of allergic inflammation. Nature. 2008;454(7203):445–54. doi:10.1038/nature07204.
103. Kay AB, Ali FR, Heaney LG, Benyahia F, Soh CP, Renz H, et al. Airway expression of calcitonin gene-related peptide in T-cell peptide-induced late asthmatic reactions in atopics. Allergy. 2007;62(5):495–503. doi:10.1111/j.1398-9995.2007.01342.x.
104. Murdoch JR, Lloyd CM. Chronic inflammation and asthma. Mutat Res. 2010;690(1–2):24–39. doi:10.1016/j.mrfmmm.2009.09.005.
105. Holgate ST. Epithelium dysfunction in asthma. J Allergy Clin Immunol. 2007;120(6):1233–1244; quiz 45-6. doi:10.1016/j.jaci.2007.10.025.
106. Gern JE, Busse WW. Relationship of viral infections to wheezing illnesses and asthma. Nat Rev Immunol. 2002;2(2):132–8. doi:10.1038/nri725.

107. Guilbert TW, Singh AM, Danov Z, Evans MD, Jackson DJ, Burton R et al. Decreased lung function after preschool wheezing rhinovirus illnesses in children at risk to develop asthma. J Allergy Clin Immunol. 2011;128(3):532–538 e1-10. doi:10.1016/j.jaci.2011.06.037.
108. Jackson DJ, Gangnon RE, Evans MD, Roberg KA, Anderson EL, Pappas TE, et al. Wheezing rhinovirus illnesses in early life predict asthma development in high-risk children. Am J Respir Crit Care Med. 2008;178(7):667–72. doi:10.1164/rccm.200802-309OC.
109. Lemanske RF Jr, Jackson DJ, Gangnon RE, Evans MD, Li Z, Shult PA, et al. Rhinovirus illnesses during infancy predict subsequent childhood wheezing. J Allergy Clin Immunol. 2005;116(3):571–7. doi:10.1016/j.jaci.2005.06.024.
110. Leung DY, Boguniewicz M, Howell MD, Nomura I, Hamid QA. New insights into atopic dermatitis. J Clin Invest. 2004;113(5):651–7. doi:10.1172/JCI21060.
111. Pawankar R, Nonaka M, Yamagishi S, Yagi T. Pathophysiologic mechanisms of chronic rhinosinusitis. Immunol Allergy Clin North Am. 2004;24(1):75–85. doi:10.1016/S0889-8561(03)00109-7.

Chapter 3
Epigenetics

3.1 Overview

Conrad Waddington coined the term "epigenetics" in 1939 [1]. It emerged by combining the Latin prefix "epi" (literally translated as "above") and the word "gene", derived from Greek, in an attempt to explain the interaction between genetics and environment. Nowadays, Epigenetics is understood as all the stable and heritable patterns and genomic functions that do not involve changes in DNA nucleotide sequence [2].

Probably one of the best examples to understand what epigenetics refers to is found in the heterogeneity of human body itself. All cells from a multicellular organism like ours, share the same genetic information; in rare cases, the DNA suffers physiological diversification processes, such as during V-D-J recombination or somatic hypermutation in B cells (discussed in Sect. 2.4.1, *Immunological basis of allergic reactions*) [3–5]. It has always been amusing that this same genetic information is able to give rise to many diverse cellular phenotypes and specific functions, ranging, for example, from a neuron (a highly specialized cell with electrically excitable properties) to cardiac myocytes (responsible for generating the electrical impulses that control the heart rate) or the gametes.

As previously stated, the term epigenetics encompasses a wide variety of mechanisms that exert long-term programs of gene expression occurring without alteration of the Watson and Crick base pair sequence [2, 6]. Among these mechanisms are methylation of DNA [7–9], the chemical modification of histones and histone variants, the ATP-dependent nucleosome remodeling complex, the Polycomb/trithorax complexes [10–13], and a variety of noncoding RNA, among which are the small interference RNA (siRNA) and microRNA (miRNA) [2, 8, 14] (see Fig. 3.1). The combined effects of these processes define the structure of chromatin at a particular locus and thus, its potential transcriptional activity.

M. Pascual and S. Roa, *Epigenetic Approaches to Allergy Research*,
SpringerBriefs in Genetics, DOI: 10.1007/978-1-4614-6366-5_3,
© The Author(s) 2013

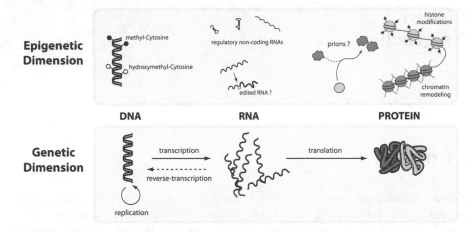

Fig. 3.1 Storage of biological information. The central dogma of molecular biology, on the *bottom*, proposes that the regular flow of genetic information goes from DNA (genes) to RNA (coding RNA), and then is translated into the proteins that carry the cellular functions. On *top* of this genetic dimension of biological information, the epigenetic dimension comprises a variety of DNA modifications, edited or noncoding RNAs, and protein elements that are key players contributing to the complexity of life and disease

3.2 DNA Methylation

The first epigenetic mechanism recognized as such was DNA methylation. DNA methylation, among all the epigenetic mechanisms, is probably the best studied and better understood epigenetic mechanism. It is based on the addition of a methyl group to the carbon 5' of a cytosine residue when followed by a guanine (Fig. 3.2), constituting what is known as a CpG dinucleotide [15]. These CpG dinucleotides are found at a lower frequency than randomly expected in the human genome [16], probably due to spontaneous deamination of methylcytosines to thymidines (CpG → TpG), making them particularly sensitive to mutations or depletions. However, in some regions known as CpG islands, these CpG dinucleotides are found at higher frequencies. In humans, the addition of methyl groups on the DNA occurs symmetrically on both DNA strands, almost exclusively on the 5' position of cytosines that are followed by guanine (CpG) [17, 18]. Other types of methylation such as those occurring in cytosines within a CpNpG or CpA context have been described in mouse embryonic stem cells and plants, but are rare in human somatic tissues [8, 17, 18].

DNA methylation can be stably transmitted to daughter cells after cell division, influencing the binding of some proteins and transcription factors to DNA, ultimately affecting locus accessibility, genetic stability, and gene expression [8]. Therefore, DNA methylation has been implicated in multiple cellular processes

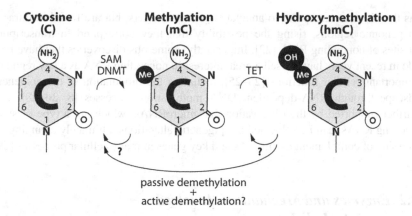

Fig. 3.2 Step-wise model of DNA methylation. Structures of cytosine (C), 5-methylcytosine (mC), and 5-hydroxymethylcytosine (hmC). The methylation of C by means of DNA methyltransferases (DNMT) and the cofactor S-adenosyl-L-methionine (SAM) is so important, that mC is sometimes considered as the 'fifth base' of the genome. Recently, hmC is gaining interest as the potential 'sixth base'. The oxidation of mC to hmC is carried out by TET enzymes, members of the 2-oxoglutarate oxygenase family. The molecular mechanisms of demethylation are still poorly understood in mammals

such as gene regulation, DNA–protein interaction, cellular differentiation, and suppression of transposable elements. Here, we briefly review our current knowledge about the ongoing studies of genomic distribution of methylation and the intensive research on the methylation/demethylation enzymatic processes. Very little is known about the impact of DNA methylation in allergic diseases, but its potential relevance can be strongly predicted from the importance of regulating methylated DNA in many other disorders and immunological mechanisms.

3.2.1 Genomic Distribution of the Methyl Mark

In certain areas of the genome, CpG dinucleotides are concentrated in CpG islands, which are defined as theoretical regions of at least 200 base pairs in length, where the content of C+G nucleotides is higher than 50 % and have an expected/observed ratio of CpG sites above 0.6 [19]. Approximately, three quarters of the total number of transcription start sites (TSS) and 88 % of active promoters in the genome are associated with CpG-rich sequences, and are subject to regulation mediated by DNA methylation [20].

Overall, only 1–2 % of the genome is composed of CpG islands or CpG-rich short sequences around the promoter region of genes, which usually are hypomethylated [21]. About half of the CpG islands in the genome are located in areas that do not correspond to annotated gene promoters, but are found in intra- or inter-genomic regions, rising the possibility that they correspond to transcription start sites of noncoding RNA [22]. Importantly, numerous discoveries that have been made in recent years have raised a great interest in noncoding RNA as a key regulator of important cellular functions [23–25], including the modulation of the epigenetic landscape through RNA-dependent DNA methylation processes [8, 26–28]. This is further supported by the observation that microRNAs, which are a type of small noncoding RNAs, can be subjected to epigenetic alterations, ultimately changing the expression of complementary mRNAs and key genes in many cellular processes [29].

3.2.2 Enzymes and Mechanisms Involved in DNA Methylation

The transfer of methyl groups to cytosines depends on the presence of donors (e.g. methionine and choline) and cofactors (e.g. folic acid, vitamin B12, or pyridoxal phosphate) to synthesize the universal donor of methyl groups, S-adenosyl-L-methionine (SAM). During methylation reactions, a methyl group is enzymatically transferred from SAM to the 5' position of a cytosine by the action of DNA methyltransferases (DNMTs) [30, 31] (Fig. 3.2). This process generates S-adenosylhomocysteine as an intermediate product, which at high concentrations can inhibit the action of DNMTs. Five DNMTs (DNMT1, DNMT2, DNMT3A, DNMT3B, and DNMT3L) have been described in humans so far, which are responsible for catalyzing the reactions of cytosine methylation by participating in maintenance processes or in *de novo* methylation processes. These enzymes have been shown to be essential during embryonic development, as demonstrated by different studies in animal models [9, 30].

It is known that DNMT1 is involved in the maintenance of methylation during cell division, as it acts on hemimethylated substrates. It is located at the replication fork during S phase of the cell cycle and methylates cytosines of the newly synthesized strand using the parental strand as a template [9, 31]. Mouse embryos lacking this enzyme exhibit a dramatic decrease in the levels of methylation, which goes down up to 70 % when compared with the expected normal global methylation levels. Furthermore, these mice die before the 11th day of gestation and histological studies have confirmed the presence of extensive cell death and reduced cell proliferation [32].

The mechanisms that control the *de novo* methylation processes are mostly unknown. It is known that DNMT3A and DNMT3B enzymes act preferentially on specific sequences and, although they have some overlapping functions, they are unable to compensate each other's functions. DNMT3A and DNMT3L are involved in the methylation of imprinted genomic regions, while DNMT3B acts mainly in repetitive regions of the genome [33]. Mice that are deficient for DNMT3A exhibit aberrant embryonic development and die within 3 to 4 weeks after birth, whereas DNMT3B-deficient mice exhibit a more severe phenotype and die before day 18 of gestation. Mice that are deficient for both enzymes, DNMT3A and DNMT3B, cannot progress during the early stages of embryonic development and die after gastrulation [34].

The DNMT2 enzyme is one of the best evolutionarily conserved methyltransferases, with homologs in species such as *Schizosaccharomyces pombe* (yeast) [35] or *Drosophila melanogaster* (fly) [36]. However, it has not been yet described any *in vitro* detectable activity of these enzymes, probably because they require cofactors for their specific activity. In this sense, other possibilities to explain this fact pointed out a preference for specific DNA sequences yet unknown [37], or the possibility that they could in fact function as RNA methyltransferases [38, 39]. Although studies in deficient mice for DNMT2 show no defect in methylation processes and have not been considered essential for development in mammals [38], studies with its counterpart in *Drosophila melanogaster* have shown that it is critical in silencing retrotransposable elements and in controlling telomeric integrity [40].

DNA methylation is closely related to other events that mediate transcriptional silencing. For example, there are proteins with different methyl binding domains (MBD) that are able to recognize and bind methylated DNA (e.g. MBD1-4), or CpG sequences (e.g. MeCP2), which are involved in the recruitment of additional protein complexes, such as histone deacetylase, polycomb, or chromatin remodeling complexes [33]. In addition to these proteins with specific binding domains for methylated DNA, it is known that methylated cytosines may also directly interact with repressor zinc finger proteins such as Kaiso, ZBTB4, or ZBTB38 [41].

Although there is no doubt about the existence of demethylation processes during development, the underlying mechanisms for a passive (involving DNA replication and "dilution" of the methyl marks) or active (involving direct enzymatic activities independently of replication) are not entirely clear [8, 42, 43] (Fig. 3.2). Recently, it has been suggested that the AID/APOBEC family of cytosine deaminases could deaminate methylated DNA, creating an intermediate DNA lesion that precedes DNA repair mechanisms and polymerization to restore the unmethylated cytosine [44–47]. This process would require repair factors such as GADD45A [48, 49], although this has been controverted [50, 51], or glycosylases such as TDG or MBD4 [43, 47, 52].

Very recently, the field of epigenetics has expanded with the discovery of exciting new DNA modifications such as hydroxymethylation (see Figs. 3.2 and 3.3). A family of enzymes, called TET proteins, has been shown to catalize the oxidation of methylcytosines to hydroxymethylcytosines (5mC => 5hmC) [53–56]. A big technological effort is being made to develop new methods to distinguish methylated and hydroxymethylated cytosines [57–60]. This has allowed us to detect 5hmC in multiple mammalian tissues [59, 61] and propose a significant epigenetic role of this modification in cellular function [55, 62, 63]. Indeed, mutations of the TET enzymes and hypermethylation have been identified in cancer [64–68] and other diseases [69], leading to the proposal that TET-mediated hydroxymethylation is important for DNA methylation fidelity in mammals by modulating active demethylation [56, 66].

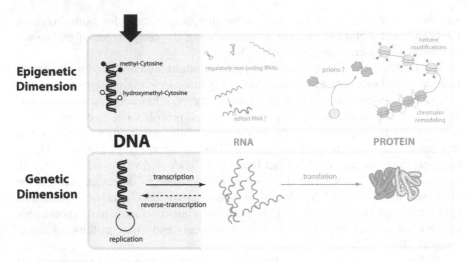

Fig. 3.3 DNA, the first epigenetic dimension. Methylation and hydroxymethylation of DNA are two major epigenetic mechanisms

Most of these enzymes and processes are present in immune cells, and even though many hematological malignancies and immune-related diseases exhibit significant epigenetic alterations [70], very little is known about DNA methylation in allergic responses (see further discussion below).

3.2.3 DNA Methylation and Disease

In mammals, epigenetic modifications play a crucial role in maintaining cellular transcription programs and organization of the architecture of DNA within the nucleus. It is known that aberrations in the epigenome may result in various diseases, some impairing normal development and others closely related to cancer development [71–73]. It is probably the field of cancer where the major advances in understanding the relationship between epigenetic aberrations and disease onset and progression have been made. However, there are other known diseases closely related to alterations in the DNA methylation patterning or recognition systems, such as the Rett syndrome. This neurological disorder primarily affects women and is due to the presence of mutations in the MeCP2 factor (methyl CpG-binding protein 2), which is able to recognize and bind to methylated CpG dinucleotides. Although the pattern of DNA methylation is not affected in this syndrome, the recognition systems are profoundly impaired, ultimately affecting the chromatin organization and regulation of gene expression [74–76].

Random silencing of one of the two X chromosomes in female mammals is one of the greatest questions still pending to be resolved, when trying to understand the transmission and stable inheritance of epigenetic information in somatic cells. In mammals, a pair of heteromorphic chromosomes (X and Y) is responsible of sex determination. Females have two X chromosomes (XX), and males have one copy of the X chromosome and one of the Y (XY). The compensation of the X chromosome gene dosage is achieved through a process of transcriptional silencing of one of the two X chromosomes in a process known as X-chromosome inactivation. This process occurs in the early stages of embryonic development in females, approximately during implantation and is stable in later development. The silencing/inactivation process is controlled by the X-inactivation centre (Xic). This region comprises about 1 Mb in length and contains several elements involved in X chromosome inactivation, including at least four genes and a remarkable untranslated RNA (Xist) [77]. Briefly, Xist starts the transcriptional silencing that occurs on the X chromosome with a physical coating of the chromosome. Soon after that, processes of chromatin remodeling, such as histone acetylation and methylation of gene promoters as well as the recruitment of the histone variant macroH2A, transform the temporary silencing induced by Xist into stable silencing of condensed chromatin [78, 79].

Another intriguing DNA methylation mediated process is genomic imprinting. Mammals are diploid organisms and have two copies of the autosomal chromosomes, one from each parent. In most cases, the expression of the alleles is independent of the maternal or paternal origin. However, some genes are expressed asymmetrically depending on its maternal or paternal inheritance, which is known as parental imprinting. Defects in the imprinting processes frequently result in neurodevelopmental and growth abnormalities, such as those seen in the case of Prader-Willy, Angelman, Beckwith-Wiedemann, and Silver-Russel syndromes [80, 81]. Regions susceptible of imprinted regulation show differential methylation between both parental alleles [82]. These areas can be maintained throughout life or may change during development, presenting a marked tissue-specific epigenomic profile [83, 84]. DNA methylation is an important mechanism linked to the imprinting phenomena, but other epigenetic marks, such as certain histone modifications have been also observed [85]. Currently, there are about 50 genes that are known to be imprinted in humans, although computational predictions estimate the existence of a larger number [86]. It is thought that other human diseases might also be related to impairment of parental imprinting, such as autism, bipolar disorders, diabetes, familial hypertrophic cardiomyopathy, schizophrenia, and even cancer [81, 87, 88]. Recently, some concerns have been raised about a possible association between epigenetic defects and imprinting disorders, which might be initiated or propagated by the use of assisted reproductive technologies [89].

The control of transposons and other mobile elements of the genome, which account for approximately 45 % of the total human genome, can also be modulated through DNA methylation mediated mechanisms. After being mistakenly considered as part of the junk DNA, repetitive elements have demonstrated to play major roles in the stabilization of the genome [90]. The repertoire of repetitive genomic elements includes Long Terminal Repeats (LTR), Long Interspersed Nuclear Elements (LINE), Short Interspersed Nuclear Elements (SINE), and transposons [91]. These mobile elements can interfere with the structure and regulation of gene expression by promoting insertions, deletions, or inversion of DNA sequences [92]. Despite the potential genetic damage that these sequences could cause, only one of every 600 mutations that occur in human germ cells is due to the action of these mobile elements [93]. Studies in fruit flies [94] and mouse embryos [95] suggest that methylation of cytosines and histone modifications are responsible for the silencing of these elements.

3.3 The RNA World

Recent discoveries have renewed interest in the RNA as a key regulator of the epigenetic landscape and a potential player in the pathogenesis of human disease (Fig. 3.4). For example, there is a growing area of research trying to understand a role for noncoding RNAs in triggering the site-specific association of DNMT enzymes, which could modulate the establishing of DNA methylation patterns [96, 97]. This process, which is known as RNA-dependent DNA methylation (RdDM) is thought to be initiated by transcription and processing of double-stranded RNA (dsRNA) that contains certain homology to the targeted genomic region [8, 98]. Other studies in mice have also suggested that maternal and paternal deposition of RNA can be a source of epigenetic information, and that transferring RNA into one-cell embryos can be sufficient to induce a pigmented phenotype and the transgenerational inheritance of different murine phenotypes [99–102].

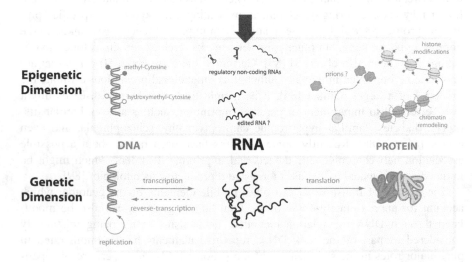

Fig. 3.4 RNA, the second epigenetic dimension. The epigenetic role of noncoding RNAs are gaining attention as modulators of the epigenetic landscape

These studies argue in favor of a role for RNA as a heritable source of epigenetic information, but could this play a role in human disease and allergy? Although much is yet to be explored, all the observations seem to favor that some of the components of complex diseases might be explained, at least in part, by variations in inherited RNAs [2]. In this sense, RNA editing, which involves the alteration of the RNA nucleotides without affecting the original DNA sequence, could contribute to the complexity of the epigenetic profile of the cell. RNA editing has been linked to tumorigenesis [103, 104], longevity [105], and some neurological disorders, including schizophrenia and depression [106, 107], but very little is known about the pathogenesis of RNA editing in human disease and virtually nothing is known about its role in allergic disease. In Sect. 3.5.3 on *How to study the epigenome*, we argue that the advent of new high-throughput sequencing technologies is opening new venues and exciting opportunities to shed light to some of the mysteries of the underexplored world of the RNA.

3.4 Protein Epigenetics

3.4.1 Unraveling a Potential 'Histone Code'

Histones are basic proteins of low molecular weight that are highly evolutionarily conserved. At the present, there are five classes of histones (H1, H2A, H2B, H3, and H4) described, which form octamers (two units of H2A, H2B, H3, and H4) that function as building blocks to package eukaryotic DNA into repeating nucleosomal units that are folded into higher-order chromatin. These units are known as nucleosomes, and between each nucleosome the linker histone (H1) can be found. The packing of the DNA around the nucleosomes defines two different states of the chromatin known as euchromatin—when the DNA is 'open' to transcription—and heterochromatin—when the DNA is 'closed' to transcription—(*Chromatin remodeling* in Fig. 3.5). In eukaryotes, the chromatin state is believed to contribute to the control of gene expression. The posttranslational modifications that occur on histones are one of the major known mechanisms of epigenetic control and include acetylation, methylation, ubiquitylation and SUMOylation of lysines, and methylation of arginines [10, 108, 109] (*Histone modifications* in Fig. 3.5). Posttranslational modifications of the histones are thought to contribute to the epigenetic control of gene expression through influencing the accessibility of protein complexes such as transcription factors, which may have an impact in many aspects of organismal development and disease onset [11].

More than 30 residues within each of the four octameric histone partners are described as targets for possible posttranslational modifications. This huge amount of modification possibilities has led to the proposal of a hypothesis called the 'histone code' [108], which postulates that "distinct histone amino-terminal modifications can generate synergistic or antagonistic interaction affinities for chromatin-associated proteins" [110]. In other words, different combinations of

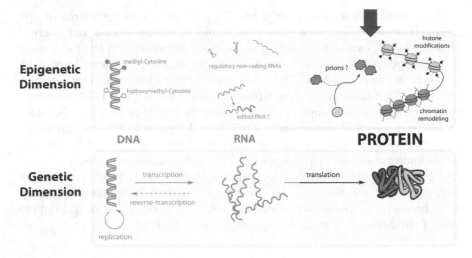

Fig. 3.5 Proteins, the third epigenetic dimension. Posttranslational modifications of the histones and chromatin remodeling are major mechanisms of epigenetic control

distinct marks in different histone residues may work as a specific code or 'protein language' prone to be recognized by different regulatory proteins. Moreover, this code may then be passed on from one cell generation to the next as an epigenetic 'memory' of transcriptional programs [13, 111, 112].

In summary, a different combination of these modifications at the amino terminus of histones has functional consequences, which affect gene activity and chromatin organization [113]. The functional consequences of these changes depend on the type of residue and the affected position. For example, among all the histone modifications, methylation (mono-, di-, or tri-) and acetylation have the most studied effects on gene regulation. These two marks often compete for the same lysine residues and they could also recruit antagonist transcription factors [2, 108, 109]. Other examples are the methylation of lysine 4 of histone H3 (H3K4) [114] and arginine 17 (H3R17) [115], which are markers of transcriptional activity, while methylation of lysine at position 9 of the histone (H3K9) is associated with transcriptional silencing [116].

3.4.2 Histone Modifications, Chromatin Remodeling, and Disease

Crosstalk and the interplay of histone modifications with other molecular pathways, such as chromatin remodeling and DNA methylation and repair has been also subject of intense research in recent years [117, 118]. The mechanisms by which cytosine methylation leads to gene silencing are closely linked to covalent modifications of chromatin remodeling proteins (such as the Polycomb (PcG) and

Trithorax (TrxG) group of proteins), histone acetyl transferases (HATs), histone deacetylases (HDACs), or methyltransferases (HMTs). All these protein complexes can not only modify histone tails, but also reorganize the chromatin architecture by altering nucleosomal positioning, ultimately modifying the epigenetic landscape and influencing the pathogenesis of many diseases [119–121].

We now know that alterations in histone "writers" or "erasers" of the histone code and deregulation of chromatin remodeling could lead to human disease. Mutations in the proteins involved in these mechanisms most often result in malignant transformation, including both solid tumors and leukemia [120, 122, 123], and have been associated with mental retardation and neurological disorders (e.g., Coffin-Lowry, Rubinstein-Taybi, and alpha thalassemia/mental retardation X-linked syndromes) [124–126]. Interestingly, it has been suggested that a prion-like epigenetic process might also mediate many neurodegenerative diseases [127, 128]. Even though this process is not fully appreciated today as a canonical form of protein epigenetics, and to the best of our knowledge has not been explored in relation to allergy, we wanted to mention it here because it is recently starting to receive increased interest. For many researchers, prionogenic proteins are emerging as an "extreme case of epigenetic inheritance" where they could play a role for cells growing under stress, for some forms of evolutionary adaptability, or for maintenance of some neuronal structural changes and persistence of long-term neurological memory [129–133].

Evidences for alterations in histone modifications that might contribute to allergic diseases are increasing, while much work remains to better understand the epigenetic basis of allergy. For example, increased HAT activity but reduced protein expression and total HDAC activity was found in bronchial specimens and peripheral blood mononuclear cells from patients with asthma when compared to healthy controls [134–136]. The authors of these studies postulate that these changes in histone acetylation could potentially result in the activation of genes that are relevant to the asthmatic phenotype. Asthma in patients who smoke or have been diagnosed with chronic obstructive pulmonary disease (COPD) are another two scenarios that have been associated with reduced activity and expression of HDAC activity [96, 97]. Interestingly, clinical and epigenetic research has revealed that corticosteroids and theophylline can attenuate the repression of HDAC activity, improving the clinical response of asthmatic patients [137–140].

Fundamental research in the epigenetic mechanisms that regulate T and B cell function, as well as other cell types involved in the allergic reaction, is also offering exciting data to support the importance of epigenetic regulation in the alterations of the immune response. For example, recent work has shown that specific histone marks, such as trimethylated histone H3 at lysine 9 (H3K9me3) and lysine 4 (H3K4me3) or acetylated at lysine 9 (H3K9ac), are important for the molecular mechanisms of class-switch recombination in B cells [141–144]. It is postulated that tightly regulated epigenetic processes are involved in the promotion of the double-strand breaks required to switch from IgM to other isotypes, and most likely to IgE too. However, much is yet to be investigated in this regard, in order to better understand the sequential mechanisms of switching from IgM to IgE [145, 146].

In the case of T cells, as another example, recent studies have demonstrated a role for HDAC1 in controlling cytokine expression levels and in modulating Th2-mediated inflammation in vivo [147].

3.5 How to Study the Epigenome

Important technical advances during last years are now allowing researchers in the epigenetic field to "look" more and more in a high-throughput manner, profiling genome-wide epigenetic marks. The use of techniques such as microarrays or massive parallel sequencing are becoming less expensive and more accessible to laboratories, providing large amounts of whole genome data, and contributing to a promising partnership between the Epigenetics and the Systems and Computational Biology fields.

There are several things to consider when designing any of these experiments:

- First, one should consider the cellular type of interest: epigenetic modifications are cell-type specific, and depending on the particular question to answer, this might be of particular interest. Moreover, to define 'abnormal', we should first define what we consider as 'normal'. The availability of source material (either DNA, chromatin, or cells) sometimes is limited and this issue may become the bottleneck of your study.
- Second, depending on the question, the experimental approach can be either 'locus-based' or 'genome-wide', and often it can start by being 'genome-wide' in a training set of specimens before validating some candidate targets in a 'locus-based' manner. The instrumental and intellectual infrastructures needed to perform these different approaches are intrinsically distinct, and they should be taken into consideration before going ahead with the project.

The following section describes the fundamental concepts and characteristics of the main techniques used to study different epigenetic marks by using both whole genome- and locus-based approaches. Numerous laboratories worldwide apply these techniques to a broad spectrum of research topics and fields, including a recently increasing body of interest in utilizing them to investigate the epigenetic aspects of allergic diseases [148–150].

3.5.1 Chromatin Immunoprecipitation

As described earlier, the eukaryote genome is packaged around an octamer of histone proteins, constituting what is known as nucleosomes, the basic unit of chromatin [151]. To investigate the interaction of chromatin and proteins such as transcription factors, cofactors, or posttranslationally modified histones, chromatin immunoprecipitation (ChIP) has been probed to be a very useful experimental technique. The first references to this technique can be found in the late 1980s, where we can find the first demonstration of the utility of ChIP experiments in the study of protein/DNA interactions [152–155]. Even though there are different versions of the original protocol, the basic steps in ChIP include the fixation of the proteins that are attached to the DNA using formaldehyde and the generation of shorter fragments by sonication. These fragments are then precipitated (or immunoprecipitated) with an antibody specific for the protein or histone modification of interest.

After precipitation, the DNA is then purified from the antibody-chromatin complex. In parallel, a duplicate of the same sample that has not been immunoprecipitated is also processed as an input control. Both samples are then analyzed for the region of interest by, for example, quantitative PCR. Thus, the ratio of enrichment by ChIP is calculated as the enrichment of the sample that has been immunoprecipitated over the one that has not gone over the process [156]. Probably, the most difficult part of the technique is to find good antibodies that show a strong affinity to the protein or the epigenetic modification of interest.

Currently, whole genome approaches combine chromatin immunoprecipitation with microarray assays (ChIP-on-chip) [157, 158] or massive parallel sequencing (ChIP-seq) [159], in order to perform high-throughput genome-wide profiling. A microarray is a solid surface that has attached a collection of microscopic probes (specific DNA sequences) that will hybridize with our sample of interest, whereas high-throughput sequencing does not use any probe. Thus, microarrays are closed designed experiments and it is necessary to take into account the analytical platform to be used. The length, density, distribution of the probes, or detection systems can significantly differ among companies (e.g. Affymetrix or Agilent). For either of these two techniques (microarrays or massive parallel sequencing), it is necessary to amplify the immunoprecipitated DNA material either by ligation-mediated PCR (LM-PCR) [160], or by linear amplification with T7 polymerase [161].

For the ChIP-seq variation of this technique, the standard ChIP protocol is followed by the massive sequencing of the immunoprecipitated material [159]. The number of reads detected in a specific genomic region is proportional to the enrichment of the targeted protein or modification within that region. So far, this technique has been carried out successfully using different histone modifications [162, 163] or transcription factors [164–166], using the massive parallel sequencing technology provided by Solexa (Illumina Inc., California, USA). More detailed information about this technology is available in Sect. 3.5.3.

3.5.2 DNA Methylation

It is increasingly clear that gene silencing through cytosine methylation of CpG dinucleotides brings many epigenetic events to cooperate for establishing a repressive environment for gene expression. Despite our knowledge on the significance of DNA methylation has increased, the exact meaning of methylation in gene bodies or intergenic regions is yet unclear; while generally speaking, methylation in the promoter regions is associated strongly with gene silencing. There is a wide spectrum of techniques described for studying methylation DNA. The basic differences in these techniques are in areas such as resolution, the ability to obtain quantitative or qualitative data, or the potential to use them in locus specific or genome-wide ways [167, 168]. Most cytosine methylation studies can be divided into three large groups (see Fig. 3.6):

1. Methods that are based on the chemical modification of DNA by the treatment with *sodium bisulfite*.
2. Methods based on the use of *restriction enzymes* sensitive to methylation.
3. Methods based on the purification of methylated DNA by *affinity pull-down and immunoprecipitation* techniques.

The methods based on the treatment of DNA with sodium bisulfite show a high level of resolution, being able to give information of each particular cytosine. This is probably the most used technique to analyze the DNA methylation levels in a particular region of interest. However, its implementation into a genome-wide level is still scarce because of bioinformatic alignment constrains, although most prognostics anticipate that it will soon be possible to use it routinely.

Other approximations, like the use of restriction enzymes or immunoprecipitation, allow the researcher to perform studies in a whole genome level, compromising the sensitivity of the technique in exchange for the possibility to profile whole genome DNA methylation levels.

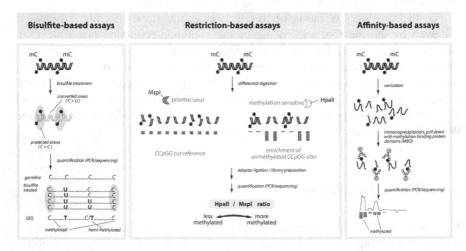

Fig. 3.6 Conceptual overview of the major technological approaches that are currently used to study DNA methylation

3.5.2.1 Bisulfite DNA Analysis

Bisulfite causes the chemical conversion of cytosine into uracil upon alkaline desulfonation. It converts 5'-methyl cytosine into uracil slower than it converts unmethylated C [169, 170], offering the possibility of a nucleotide-level resolution profiling of DNA methylation (Fig. 3.6). During PCR amplification, uracil transformed cytosines and unconverted cytosines will be amplified as thymidine (T) and cytosine (C), respectively, thus offering the possibility to compare the "original sequence" with the new treated and amplified sequence.

One way to compare bisulfite-treated DNA from different samples is to utilize *methylation microarray chips*, where labeled DNA is hybridized to probes fixed onto a solid surface. Nowadays, probably the most widely used microarray platforms to assess DNA methylation levels are the *HumanMethylation27 BeadChip* and the *Infinium HumanMethylation450 BeadChip*, both provided by Illumina (http://www.illumina.com). They are very reliable, well-validated, and cost-effective platforms, and the quantity of DNA they require do not exceed 1 microgram. Both designs give quantitative information at single-nucleotide resolution. However, the *HumanMethylation27 BeadChip* interrogates 27,578 CpG loci corresponding to more than 14,000 genes, whereas the *Infinium HumanMethylation450 BeadChip* covers more than 485,000 methylation sites per sample, including promoter, 5'UTR, first exon, gene body, and 3'UTR, CpG islands outside of coding regions, and miRNA promoter regions.

There are several techniques for the quantification of methylation levels in a particular region of interest. Maybe one of the most used and most laborious is the *cloning and subsequent Sanger sequencing* of the amplified fragment. To achieve statistical power, between 10 and 20 different colonies are recommended to be sequenced in search for a 'bisulfite signature' of T residues instead of the germline unmethylated C residues. On the contrary, as shown in Fig. 3.6, methylated Cs (mC) are not converted to uracil by the sodium bisulfite reaction (i.e. by the sulfonation reaction, hydrolytic deamination, and final alkaline-mediated desulfonation), and will remain as C residues during the Sanger sequencing.

Pyrosequencing [171] has also been adapted to the study of DNA methylation. After PCR amplification of bisulfite-treated DNA, the fragment of interest is pyrosequenced, while amplified by using a luminescent signal that is generated after the incorporation of each nucleotide. Usually, one of the primers used during the PCR amplification is biotinylated, so the fragment of interest is immobilized on streptavidin beads, where the sequencing occurs as single stranded DNA. The degree of methylation at each CpG position is determined by the T/C ratio with higher quantitative resolution than traditional Sanger sequencing. One of the main advantages of this method is its relatively simplicity, including protocols that can provide information for up to 96 DNA samples in approximately 4 h [172–175]. However, this technique is limited to relatively small 25–30 bp long fragments.

The methylation-specific PCR or MSP (for its acronym Methylation Specific PCR) [176], also uses bisulfite-treated DNA as template. This method is able to interrogate the methylation status of virtually any group of CpG sites within a

CpG island. Two different sets of primers targeting the same region are used, one pair of them being designed for methylated DNA, whereas the other pair is specifically designed to amplify the unmethylated DNA. Therefore, the design of the primers is critical, and relays on the notion that unmethylated regions are usually enriched in converted Uracile residues (or Thymidine after PCR). Numerous algorithms have been developed to address the criteria and optimization rules in primer design for this kind of approaches (e.g. MethPrimer, http://www.urogene.org/me thprimer/index1.html). The PCR products are then visualized in an agarose gel. Although MSP is a quick and effective method, it should be noted that this technique is qualitative rather than quantitative, since it does not give any percentage of methylation of any given CpG dinucleotide, but an overall view of the methylation level of the amplified fragment.

In *COBRA* (for its acronym *CO*mbined *B*isulphite *R*estriction *A*nalysis) [177], the DNA fragment of interest is bisulfite treated before being amplified by PCR. After amplification, the amplicon is digested with a restriction enzyme like BstUI (CGCG), that will be unable to cut if the restriction site was not methylated, and thus the sequence was prone to be changed. The result of the digestion is then analyzed by Southern blot or any other quantitative method to calculate the ratio of methylated/unmethylated fragments. Besides being a fast and simple quantitative method, its use is limited to regions that can be targeted by such restriction enzymes.

MassARRAY is another quantitative technique used to study the levels of methylation of each CpG nucleotide within a given region. It is based on MALDI/TOF mass spectrometry for quantification [178]. After bisulfite conversion of DNA and PCR amplification, the reverse strand of the amplicon is transcribed into a single strand RNA molecule, followed by a base specific RNAse cleavage. This cleavage will result in a different pattern of fragments, depending on the methylation status of the original CpG dinucleotide. The fragment mass is then determined by MALDI/TOF mass spectrometry, and the cleavage products are automatically and quantitatively measured with the MassARRAY® EpiTYPER™ system and software. This technique is fast and simple, and as the other PCR-based techniques, it needs a relatively small amount of DNA. It can interrogate larger fragments than those analyzed by pyrosequencing (around 200–300 bp instead of 20–30 bp of the pyrosequencing). Unlike other more affordable techniques described previously that involve more standard molecular biology techniques, the high cost of the equipment needed to run MassARRAY experiments explains why this technique is less widespread among the scientific community. On the other hand, it is becoming a locus-based method of choice for many laboratories, primarily because of its quantitative sensitivity, accuracy, and potential for high-throughput implementation.

3.5.2.2 Use of Restriction Enzymes

Differential methylation hybridization (DMH) [179], is a high-throughput technique based on the use of methylation sensitive restriction enzymes. Total genomic DNA is first sonicated into 400–500 bp long fragments and then ligated using linkers. These ligated DNA fragments are then incubated with two restriction enzymes like *Hpa*II and *Hin*P1I, which have the recognition cutting sites CCGG and CGCG, respectively. After digestion, a PCR is carried out that will only amplify intact fragments. Thus, if a given fragment is cleaved by at least one of the enzymes, it will not hybridize properly onto a microarray that contain probes spanning the 27,800 CpG islands annotated in the UCSC Genome Browser [180]. This technique can detect differentially methylated CpG-rich regions between two experimental conditions, using one as 'reference'.

Another widely used method based on the use of restriction enzymes is the HELP assay, from its acronym *Hpa*II tiny fragment *E*nrichment by *L*igation-mediated *P*CR [181, 182]. High molecular weight genomic DNA is digested in parallel with the enzyme *Hpa*II and its isoschizomer *Msp*I. Both enzymes target the same sequence (CCGG), but while *Hpa*II is methylation sensitive, *Msp*I can always cut both CCGG and CmCGG sequences regarding its methylation status. The result of each digestion is ligated to a set of primers, PCR amplified (this process is known as ligation-mediated PCR, LM-PCR) and size-selected. *Hpa*II and *Msp*I digested LM-PCRs are then labeled with different fluorescent dyes and hybridized to a custom microarray, designed specifically to represent all the genomic regions predicted to contain CCGG sites. HELP assay is a semiquantitative technique that allows researchers to interrogate the methylation levels not only of promoters and gene bodies, but also of intergenic regions.

A new version of HELP assay has been recently published, known as HELP-tagging [183, 184]. In this version, the authors take advantage of the new technologies available today (like deep sequencing), successfully adapting their protocol to this new technology. Similar to the standard HELP assay, in HELP-tagging DNA samples are digested with *Hpa*II and ligated to a customized Illumina adapter with a complementary cohesive end that also contains an *Eco*P15I site. This allows one to cut the adjacent sequence 27 bp away from the site of digestion, polish that end, and prepare libraries for PCR by ligating other Illumina adapters containing 'bar codes', which enables multiplexing on deep sequencing platforms (e.g. Illumina HiSeq2000 platform). The CCGG and *Eco*P15I motifs permit removal of spurious sequences and provides ~100 bp long fragments ideal for current deep sequencing technologies. Another implementation of this assay is that prior to library preparation and sequencing, T7 polymerase-mediated in vitro transcription and qRT-PCR are performed, which allow the filtering of contaminating products and single-adapter products. Following sequencing, a Wiki-based Automated Sequence Processor (WASP) system supported by the Epigenomics Shared Facility of the Albert Einstein College of Medicine (NY, USA) (http://wasp.einstein.yu.edu/), provides high-performance computing resources and analytical tools [184].

3.5.2.3   Purification by Affinity

One of the ways to enrich the portion of methylated DNA within the genome is based on isolating the methylated fragments from those that are unmethylated.

There are several technologies developed to do that, such as those that use purification columns with Methyl Binding Domains (MBDs), or those that use monoclonal antibodies to immunoprecipitate fragments enriched with methylated cytosines (MeDIP). The purified fraction of methylated DNA can then be analyzed in a locus-specific manner, or can be input to high-throughput methods such as high-resolution DNA microarrays or next-generation sequencing.

The major concern with these techniques is that in practice, they are dependent on the density of methylated cytosines in a given region. In general, the genomes of mammals have a low density of CpG residues, except for those areas where they are highly concentrated and form CpG islands, which are generally unmethylated [185]. Therefore, genome areas enriched in CpG sites and CpG islands could be prone to more efficient pull downs, which may introduce unpredicted biases. This is one of the main reasons that is currently pushing researchers to rather favor bisulfite-mediated or restriction enzyme-mediated approaches to study DNA methylation.

3.5.3 New Technologies

The Human Genome Project (HGP) was a 13-year project, completed in 2003, with an estimated investment of $3.8 billion dollars (http://www.ornl. gov/sci/techresources/Human_Genome/home.shtml). Nowadays, with the advent of the massive parallel sequencing (or deep sequencing) technology, we are approaching to a new era of high-throughput research. As a result, a new challenge has been proposed: "the $1,000 Genome" project, where the main goals are to drop the expenses of full genome sequencing of an individual (or patient) to $1,000 and to improved performance so this could be done in the matter of days.

As it was mentioned before, most of the technical approaches to study the epigenome that we have mentioned in this chapter can be implemented to be used with deep sequencing technologies: BS (by sequencing bisulfite-treated DNA), HELP-tagging (by utilizing methylation sensitive restriction enzymes), ChIP-seq (by utilizing immunoprecipitation), MeDIP-Seq (by utilizing affinity pull downs), Whole Genome Shotgun sequencing (by sequencing restriction-generated longer fragments in an attempt to overcome the assembly problems of other bisulfite-related methods), or RNA-seq (by sequencing coding and noncoding regulatory RNAs). One of the techniques that is gaining momentum is an approach termed reduced representation bisulfite sequencing (RRBS) [186–188]. This genome-wide assay also requires the use of bisulfite to distinguish unmethylated from methylated DNA, but enriches first CpG-rich parts of the genome. This is achieved by cutting the untreated DNA with the methylation-insensitive *Msp*I enzyme, which generates short fragments that contain CpG nucleotides at the ends and, after size selection and bisulfite treatment, allows to reduce the amount of sequencing required.

Sequencing platforms such as the 454 FLX massive parallel sequencing provided by Roche (Life Sciences, Roche Applied Science, Indianapolis, USA), or

Illumina GA (Illumina Inc., California, USA) technologies have allowed and revolutionized the way to perform whole genome sequencing techniques. While the later is able to generate ~40 million reads of 100 bp in length, which accounts to more than one billion bp sequenced, the former generates ~400,000 reads of more than 250 bp, giving a total of over 100 million bases sequenced. Massive parallel sequencing from Illumina is based on massive sequencing of clusters, generated from approximately 1000 identical DNA fragments, each of them generated from a single DNA molecule. This reaction is carried out in a glass platform that can accommodate up to 50 million clusters.

Probably, one of the main limitations of this technology is its cost. Although if you compare the total price for the entire sequencing process with the potential price of a Sanger sequencing approach, the first might seem cheaper. In reality, however, these methods are still available to only a few laboratories. Moreover, the computational analysis of the data requires complex processing, which is exponentially increasing the requirement of cooperation with computational and systems biologists. Although the development and implementation of analytical tools and algorithms is challenging and might become a limiting factor for many laboratories, numerous international initiatives are constantly arising to study epigenetics in a more integrated and cooperative manner, providing access to many analytical resources and pipelines. Some of them are the ENCODE (ENCyclopedia Of DNA Elements, http://www.genome.gov/10005107), DAnCER (Disease Annotated Chromatin Epigenetic Resource, http://wodaklab.org/dancer), HEROIC (High- throughput Epigenetic Regulatory Organization in Chromatin, http://www.heroic-ip.eu), the HEP (Human Epigenome Project, http://www.epigenome.org), or the NIH Common Fund's Epigenomics Program (http://commonfund.nih.gov/epigenomics).

Indeed, the field of asthma and allergy, among others, would highly benefit from the fast development, the increase in accessibility to analytical pipelines and resources, and the reduction of costs of this promising area of massive parallel sequencing, in order to perform epigenetic research in allergy-related specimens.

References

1. Waddington CH. Preliminary notes on the development of the wings in normal and mutant strains of drosophila. Proc Natl Acad Sci USA. 1939;25(7):299–307.
2. Chahwan R, Wontakal SN, Roa S. The multidimensional nature of epigenetic information and its role in disease. Discov Med. 2011;11(58):233–43.
3. Peled JU, Kuang FL, Iglesias-Ussel MD, Roa S, Kalis SL, Goodman MF, et al. The biochemistry of somatic hypermutation. Annu Rev Immunol. 2008;26:481–511.
4. Kato L, Stanlie A, Begum NA, Kobayashi M, Aida M, Honjo T. An evolutionary view of the mechanism for immune and genome diversity. J Immunol. 2012;188(8):3559–66. doi:10.4049/jimmunol.1102397.
5. Lieber MR. The mechanism of double-strand DNA break repair by the nonhomologous DNA end-joining pathway. Annu Rev Biochem. 2010;79:181–211. doi:10.1146/annurev.biochem.052308.093131.

6. Delcuve GP, Rastegar M, Davie JR. Epigenetic control. J Cell Physiol. 2009;219(2):243–50. doi:10.1002/jcp.21678.
7. Jones PA. Functions of DNA methylation: islands, start sites, gene bodies and beyond. Nat Rev. 2012;. doi:10.1038/nrg3230.
8. Law JA, Jacobsen SE. Establishing, maintaining and modifying DNA methylation patterns in plants and animals. Nat Rev. 2010;11(3):204–20. doi:nrg2719 [pii]10.1038/nrg2719.
9. Klose RJ, Bird AP. Genomic DNA methylation: the mark and its mediators. Trends Biochem Sci. 2006;31(2):89–97. doi:10.1016/j.tibs.2005.12.008.
10. Bartke T, Vermeulen M, Xhemalce B, Robson SC, Mann M, Kouzarides T. Nucleosome-interacting proteins regulated by DNA and histone methylation. Cell. 2010;143(3):470–84. doi:S0092-8674(10)01182-7 [pii]10.1016/j.cell.2010.10.012.
11. Pedersen MT, Helin K. Histone demethylases in development and disease. Trends Cell Biol. 2010;20(11):662–71. doi:S0962-8924(10)00179-0 [pii]10.1016/j.tcb.2010.08.011.
12. Mellor J. The dynamics of chromatin remodeling at promoters. Mol Cell. 2005;19(16039585):147–57.
13. Bird A. Perceptions of epigenetics. Nature. 2007;447(7143):396–8. doi:nature05913 [pii]10.1038/nature05913.
14. Bourc'his D, Voinnet O. A small-RNA perspective on gametogenesis, fertilization, and early zygotic development. Science (New York, NY. 2010;330(6004):617–22. doi:330/6004/617 [pii]10.1126/science.1194776.
15. Bird A. DNA methylation patterns and epigenetic memory. Genes Dev. 2002;16(1):6–21. doi:10.1101/gad.947102.
16. Bird AP. CpG-rich islands and the function of DNA methylation. Nature. 1986;321(6067):209–13. doi:10.1038/321209a0.
17. Pelizzola M, Ecker JR. The DNA methylome. FEBS Lett. 2011;585(13):1994–2000. doi:10.1016/j.febslet.2010.10.061.
18. Lister R, Pelizzola M, Dowen RH, Hawkins RD, Hon G, Tonti-Filippini J, et al. Human DNA methylomes at base resolution show widespread epigenomic differences. Nature. 2009;462(7271):315–22. doi:10.1038/nature08514.
19. Glass JL, Thompson RF, Khulan B, Figueroa ME, Olivier EN, Oakley EJ et al. CG dinucleotide clustering is a species-specific property of the genome. Nucleic Acid Res. 2007;35(20):6798–807. doi:gkm489 [pii]10.1093/nar/gkm489.
20. Feltus FA, Lee EK, Costello JF, Plass C, Vertino PM. DNA motifs associated with aberrant CpG island methylation. Genomics. 2006;87(5):572–9. doi:10.1016/j.ygeno.2005.12.016.
21. Suzuki MM, Bird A. DNA methylation landscapes: provocative insights from epigenomics. Nat Rev. 2008;9(6):465–76. doi:10.1038/nrg2341.
22. Illingworth R, Kerr A, Desousa D, Jorgensen H, Ellis P, Stalker J, et al. A novel CpG island set identifies tissue-specific methylation at developmental gene loci. PLoS Biol. 2008;6(1):e22. doi:10.1371/journal.pbio.0060022.
23. Huttenhofer A, Schattner P, Polacek N. Non-coding RNAs: hope or hype? Trends Genet. 2005;21(5):289–97. doi:10.1016/j.tig.2005.03.007.
24. Eddy SR, Non-coding RNA. Genes and the modern RNA world. Nat Rev. 2001;2(12):919–29. doi:10.1038/35103511.
25. Rinn JL, Chang HY. Genome regulation by long noncoding RNAs. Annu Rev Biochem. 2012;81:145–66. doi:10.1146/annurev-biochem-051410-092902.
26. Brennecke J, Malone CD, Aravin AA, Sachidanandam R, Stark A, Hannon GJ. An epigenetic role for maternally inherited piRNAs in transposon silencing. Science. 2008;322(5906):1387–92. doi:322/5906/1387 [pii]10.1126/science.1165171.
27. Aravin AA, Sachidanandam R, Bourc'his D, Schaefer C, Pezic D, Toth KF et al. A piRNA pathway primed by individual transposons is linked to de novo DNA methylation in mice. Mol Cell. 2008;31(6):785–99. doi:S1097-2765(08)00619-9 [pii]10.1016/j.molcel.2008.09.003.
28. Kuramochi-Miyagawa S, Watanabe T, Gotoh K, Totoki Y, Toyoda A, Ikawa M et al. DNA methylation of retrotransposon genes is regulated by Piwi family members MILI and

MIWI2 in murine fetal testes. Genes Dev. 2008;22(7):908–17. doi:22/7/908[pii]10.1101/gad.1640708.

29. Guil S, Esteller M. DNA methylomes, histone codes and miRNAs: tying it all together. Int J Biochem Cell Biol. 2009;41(1):87–95. doi:10.1016/j.biocel.2008.09.005.

30. Malygin EG, Hattman S. DNA methyltransferases: mechanistic models derived from kinetic analysis. Crit Rev Biochem Mol Biol. 2012;47(2):97–193. doi:10.3109/10409238.2011.620942.

31. Cedar H, Bergman Y. Programming of DNA methylation patterns. Annu Rev Biochem. 2012;81:97–117. doi:10.1146/annurev-biochem-052610-091920.

32. Li E, Bestor TH, Jaenisch R. Targeted mutation of the DNA methyltransferase gene results in embryonic lethality. Cell. 1992;69(6):915–26.

33. Tost J. DNA methylation: an introduction to the biology and the disease-associated changes of a promising biomarker. Mol Biotechnol. 2010;44(1):71–81. doi:10.1007/s12033-009-9216-2.

34. Okano M, Bell DW, Haber DA, Li E. DNA methyltransferases Dnmt3a and Dnmt3b are essential for de novo methylation and mammalian development. Cell. 1999;99(3):247–57.

35. Yoder JA, Bestor TH. A candidate mammalian DNA methyltransferase related to pmt1p of fission yeast. Hum Mol Genet. 1998;7(2):279–84.

36. Tweedie S, Ng HH, Barlow AL, Turner BM, Hendrich B, Bird A. Vestiges of a DNA methylation system in Drosophila melanogaster? Nat Genet. 1999;23(4):389–90. doi:10.1038/70490.

37. Hermann A, Schmitt S, Jeltsch A. The human Dnmt2 has residual DNA-(cytosine-C5) methyltransferase activity. The J Biol Chem. 2003;278(34):31717–21. doi:10.1074/jbc.M305448200.

38. Okano M, Xie S, Li E. Dnmt2 is not required for de novo and maintenance methylation of viral DNA in embryonic stem cells. Nucleic Acids Res. 1998;26(11):2536–40.

39. Schaefer M, Pollex T, Hanna K, Tuorto F, Meusburger M, Helm M, et al. RNA methylation by Dnmt2 protects transfer RNAs against stress-induced cleavage. Genes Dev. 2010;24(15):1590–5. doi:10.1101/gad.586710.

40. Phalke S, Nickel O, Walluscheck D, Hortig F, Onorati MC, Reuter G. Retrotransposon silencing and telomere integrity in somatic cells of Drosophila depends on the cytosine-5 methyltransferase DNMT2. Nat Genet. 2009;41(6):696–702. doi:10.1038/ng.360.

41. Filion GJ, Zhenilo S, Salozhin S, Yamada D, Prokhortchouk E, Defossez PA. A family of human zinc finger proteins that bind methylated DNA and repress transcription. Mol Cell Biol. 2006;26(1):169–81. doi:10.1128/MCB.26.1.169-181.2006.

42. Wu SC, Zhang Y. Active DNA demethylation: many roads lead to Rome. Nat Rev Mol Cell Biol. 2010. doi:nrm2950[pii]10.1038/nrm2950.

43. Zhu JK. Active DNA demethylation mediated by DNA glycosylases. Annu Rev Genet. 2009;43:143–66. doi:10.1146/annurev-genet-102108-134205.

44. Bhutani N, Brady JJ, Damian M, Sacco A, Corbel SY, Blau HM. Reprogramming towards pluripotency requires AID-dependent DNA demethylation. Nature. 2009;463(7284):1042–7. doi:nature08752 [pii]10.1038/nature08752.

45. Morgan HD, Dean W, Coker HA, Reik W, Petersen-Mahrt SK. Activation-induced cytidine deaminase deaminates 5-methylcytosine in DNA and is expressed in pluripotent tissues: implications for epigenetic reprogramming. The J Biol Chem. 2004;279(50):52353–60.

46. Popp C, Dean W, Feng S, Cokus SJ, Andrews S, Pellegrini M et al. Genome-wide erasure of DNA methylation in mouse primordial germ cells is affected by AID deficiency. Nature. 2010;463(7284):1101–5. doi:nature08829 [pii]10.1038/nature08829.

47. Rai K, Huggins IJ, James SR, Karpf AR, Jones DA, Cairns BR. DNA demethylation in zebrafish involves the coupling of a deaminase, a glycosylase, and gadd45. Cell. 2008;135(7):1201–12. doi:S0092-8674(08)01517-1 [pii]10.1016/j.cell.2008.11.042.

48. Sytnikova YA, Kubarenko AV, Schafer A, Weber AN, Niehrs C. Gadd45a is an RNA binding protein and is localized in nuclear speckles. PLoS ONE. 2011;6(1):e14500. doi:10.1371/journal.pone.0014500.

49. Barreto G, Schafer A, Marhold J, Stach D, Swaminathan SK, Handa V et al. Gadd45a promotes epigenetic gene activation by repair-mediated DNA demethylation. Nature. 2007;445(7128):671–5. doi:nature05515 [pii]10.1038/nature05515.
50. Engel N, Tront JS, Erinle T, Nguyen N, Latham KE, Sapienza C, et al. Conserved DNA methylation in Gadd45a(-/-) mice. Epigenetics Official J DNA Methylation Soc. 2009;4(2):98–9.
51. Jin SG, Guo C, Pfeifer GP. GADD45A does not promote DNA demethylation. PLoS Genet. 2008;4(3):e1000013. doi:10.1371/journal.pgen.1000013.
52. Cortellino S, Xu J, Sannai M, Moore R, Caretti E, Cigliano A, et al. Thymine DNA glycosylase is essential for active DNA demethylation by linked deamination-base excision repair. Cell. 2011;146(1):67–79. doi:10.1016/j.cell.2011.06.020.
53. Ito S, D'Alessio AC, Taranova OV, Hong K, Sowers LC, Zhang Y. Role of Tet proteins in 5mC to 5hmC conversion, ES-cell self-renewal and inner cell mass specification. Nature. 2010;466(7310):1129–33. doi:nature09303 [pii]10.1038/nature09303.
54. Ko M, Huang Y, Jankowska AM, Pape UJ, Tahiliani M, Bandukwala HS et al. Impaired hydroxylation of 5-methylcytosine in myeloid cancers with mutant TET2. Nature. 2010. doi:nature09586 [pii]10.1038/nature09586.
55. Tahiliani M, Koh KP, Shen Y, Pastor WA, Bandukwala H, Brudno Y et al. Conversion of 5-methylcytosine to 5-hydroxymethylcytosine in mammalian DNA by MLL partner TET1. Science. 2009;324(5929):930–5. doi:1170116 [pii]10.1126/science.1170116.
56. Guo JU, Su Y, Zhong C, Ming GL, Song H. Hydroxylation of 5-Methylcytosine by TET1 promotes active DNA demethylation in the adult brain. Cell. 2011;145(3):423–34. doi:10.1016/j.cell.2011.03.022.
57. Booth MJ, Branco MR, Ficz G, Oxley D, Krueger F, Reik W et al. Quantitative sequencing of 5-Methylcytosine and 5-Hydroxymethylcytosine at single-base resolution. Science. 2012. doi:10.1126/science.1220671.
58. Song CX, Szulwach KE, Fu Y, Dai Q, Yi C, Li X, et al. Selective chemical labeling reveals the genome-wide distribution of 5-hydroxymethylcytosine. Nat Biotechnol. 2011;29(1):68–72. doi:10.1038/nbt.1732.
59. Szwagierczak A, Bultmann S, Schmidt CS, Spada F, Leonhardt H. Sensitive enzymatic quantification of 5-hydroxymethylcytosine in genomic DNA. Nucleic Acids Res. 2010;38(19):e181. doi:gkq684 [pii]10.1093/nar/gkq684.
60. Gupta R, Nagarajan A, Wajapeyee N. Advances in genome-wide DNA methylation analysis. Biotechniques. 2010;49(4):iii–xi. doi:000113493 [pii]10.2144/000113493.
61. Globisch D, Munzel M, Muller M, Michalakis S, Wagner M, Koch S, et al. Tissue distribution of 5-hydroxymethylcytosine and search for active demethylation intermediates. PLoS ONE. 2010;5(12):e15367. doi:10.1371/journal.pone.0015367.
62. Branco MR, Ficz G, Reik W. Uncovering the role of 5-hydroxymethylcytosine in the epigenome. Nat Rev. 2012;13(1):7–13. doi:10.1038/nrg3080.
63. Kriaucionis S, Heintz N. The nuclear DNA base 5-hydroxymethylcytosine is present in Purkinje neurons and the brain. Science. 2009;324(5929):929–30. doi:1169786 [pii]10.1126/science.1169786.
64. Saint-Martin C, Leroy G, Delhommeau F, Panelatti G, Dupont S, James C, et al. Analysis of the ten-eleven translocation 2 (TET2) gene in familial myeloproliferative neoplasms. Blood. 2009;114(8):1628–32. doi:10.1182/blood-2009-01-197525.
65. Delhommeau F, Dupont S, Della Valle V, James C, Trannoy S, Masse A, et al. Mutation in TET2 in myeloid cancers. N Engl J Med. 2009;360(22):2289–301. doi:10.1056/NEJMoa0810069.
66. Williams K, Christensen J, Helin K. DNA methylation: TET proteins-guardians of CpG islands? EMBO Rep. 2012;13(1):28–35. doi:10.1038/embor.2011.233.
67. Jankowska AM, Szpurka H, Tiu RV, Makishima H, Afable M, Huh J, et al. Loss of heterozygosity 4q24 and TET2 mutations associated with myelodysplastic/myeloproliferative neoplasms. Blood. 2009;113(25):6403–10. doi:10.1182/blood-2009-02-205690.

68. Langemeijer SM, Kuiper RP, Berends M, Knops R, Aslanyan MG, Massop M, et al. Acquired mutations in TET2 are common in myelodysplastic syndromes. Nat Genet. 2009;41(7):838–42. doi:10.1038/ng.391.

69. Tefferi A, Pardanani A, Lim KH, Abdel-Wahab O, Lasho TL, Patel J, et al. TET2 mutations and their clinical correlates in polycythemia vera, essential thrombocythemia and myelofibrosis. Leukemia. 2009;23(5):905–11. doi:10.1038/leu.2009.47.

70. Rodriguez-Cortez VC, Hernando H, de la Rica L, Vento R, Ballestar E. Epigenomic deregulation in the immune system. Epigenomics. 2011;3(6):697–713. doi:10.2217/epi.11.99.

71. Esteller M. Cancer epigenetics for the twenty first century: what's next? Genes Cancer. 2011;2(6):604–6. doi:10.1177/1947601911423096.

72. Rodriguez-Paredes M, Esteller M. Cancer epigenetics reaches mainstream oncology. Nat Med. 2011;17(3):330–9. doi:10.1038/nm.2305.

73. Esteller M. Epigenetics in cancer. N Engl J Med. 2008;358(11):1148–59. doi:10.1056/NEJMra072067.

74. Amir RE, Van den Veyver IB, Wan M, Tran CQ, Francke U, Zoghbi HY. Rett syndrome is caused by mutations in X-linked MECP2, encoding methyl-CpG-binding protein 2. Nat Genet. 1999;23(2):185–8. doi:10.1038/13810.

75. Samaco RC, Neul JL. Complexities of Rett syndrome and MeCP2. J Neurosci. 2011;31(22):7951–9. doi:10.1523/JNEUROSCI.0169-11.2011.

76. Zachariah RM, Rastegar M. Linking epigenetics to human disease and Rett syndrome: the emerging novel and challenging concepts in MeCP2 research. Neural plasticity. 2012;2012:415825. doi:10.1155/2012/415825.

77. Avner P, Heard E. X-chromosome inactivation: counting, choice and initiation. Nat Rev. 2001;2(1):59–67. doi:10.1038/35047580.

78. Morey C, Navarro P, Debrand E, Avner P, Rougeulle C, Clerc P. The region 3' to Xist mediates X chromosome counting and H3 Lys-4 dimethylation within the Xist gene. EMBO J. 2004;23(3):594–604. doi:10.1038/sj.emboj.7600071.

79. Chow JC, Brown CJ. Forming facultative heterochromatin: silencing of an X chromosome in mammalian females. Cell Mol Life Sci. 2003;60(12):2586–603. doi:10.1007/s00018-003-3121-9.

80. Sleutels F, Barlow DP. The origins of genomic imprinting in mammals. Adv Genet. 2002;46:119–63.

81. Falls JG, Pulford DJ, Wylie AA, Jirtle RL. Genomic imprinting: implications for human disease. Am J Pathol. 1999;154(3):635–47.

82. Spahn L, Barlow DP. An ICE pattern crystallizes. Nat Genet. 2003;35(1):11–2. doi:10.1038/ng0903-11.

83. Stoger R, Kubicka P, Liu CG, Kafri T, Razin A, Cedar H, et al. Maternal-specific methylation of the imprinted mouse Igf2r locus identifies the expressed locus as carrying the imprinting signal. Cell. 1993;73(1):61–71.

84. Feil R, Walter J, Allen ND, Reik W. Developmental control of allelic methylation in the imprinted mouse Igf2 and H19 genes. Development. 1994;120(10):2933–43.

85. Xin Z, Tachibana M, Guggiari M, Heard E, Shinkai Y, Wagstaff J. Role of histone methyltransferase G9a in CpG methylation of the Prader-Willi syndrome imprinting center. The J Biol Chem. 2003;278(17):14996–5000. doi:10.1074/jbc.M211753200M211753200 [pii].

86. Luedi PP, Dietrich FS, Weidman JR, Bosko JM, Jirtle RL, Hartemink AJ. Computational and experimental identification of novel human imprinted genes. Genome Res. 2007;17(12):1723–30. doi:10.1101/gr.6584707.

87. Jirtle RL. Genomic imprinting and cancer. Exp Cell Res. 1999;248(1):18–24. doi:S0014-4827(99)94453-1 [pii]10.1006/excr.1999.4453.

88. Morison IM, Reeve AE. A catalogue of imprinted genes and parent-of-origin effects in humans and animals. Hum Mol Genet. 1998;7(10):1599–609. doi:ddb178 [pii].

89. Lawrence LT, Moley KH. Epigenetics and assisted reproductive technologies: human imprinting syndromes. Semin Reprod Med. 2008;26(2):143–52. doi:10.1055/s-2008-1042953.

90. Lander ES, Linton LM, Birren B, Nusbaum C, Zody MC, Baldwin J, et al. Initial sequencing and analysis of the human genome. Nature. 2001;409(6822):860–921. doi:10.1038/35057062.
91. Cordaux R, Batzer MA. The impact of retrotransposons on human genome evolution. Nat Rev. 2009;10(10):691–703. doi:10.1038/nrg2640.
92. Kim JK, Samaranayake M, Pradhan S. Epigenetic mechanisms in mammals. Cell Mol Life Sci. 2009;66(4):596–612. doi:10.1007/s00018-008-8432-4.
93. Kazazian HH Jr. An estimated frequency of endogenous insertional mutations in humans. Nat Genet. 1999;22(2):130. doi:10.1038/9638.
94. Rae PM, Steele RE. Absence of cytosine methylation at C-C-G-G and G-C-G-C sites in the rDNA coding regions and intervening sequences of Drosophila and the rDNA of other insects. Nucleic Acids Res. 1979;6(9):2987–95.
95. Walsh CP, Chaillet JR, Bestor TH. Transcription of IAP endogenous retroviruses is constrained by cytosine methylation. Nat Genet. 1998;20(2):116–7. doi:10.1038/2413.
96. Ito K, Ito M, Elliott WM, Cosio B, Caramori G, Kon OM, et al. Decreased histone deacetylase activity in chronic obstructive pulmonary disease. N Engl J Med. 2005;352(19):1967–76. doi:10.1056/NEJMoa041892.
97. Chaudhuri R, Livingston E, McMahon AD, Thomson L, Borland W, Thomson NC. Cigarette smoking impairs the therapeutic response to oral corticosteroids in chronic asthma. Am J Respir Crit Care Med. 2003;168(11):1308–11. doi:10.1164/rccm.200304-503OC.
98. Mette MF, Aufsatz W, van der Winden J, Matzke MA, Matzke AJ. Transcriptional silencing and promoter methylation triggered by double-stranded RNA. EMBO J. 2000;19(19):5194–201. doi:10.1093/emboj/19.19.5194.
99. Wagner KD, Wagner N, Ghanbarian H, Grandjean V, Gounon P, Cuzin F et al. RNA induction and inheritance of epigenetic cardiac hypertrophy in the mouse. Dev Cell. 2008;14(6):962–9. doi:S1534-5807(08)00119-6 [pii]10.1016/j.devcel.2008.03.009.
100. Rassoulzadegan M, Grandjean V, Gounon P, Vincent S, Gillot I, Cuzin F. RNA-mediated non-mendelian inheritance of an epigenetic change in the mouse. Nature. 2006;441(7092):469–74. doi:nature04674 [pii]10.1038/nature04674.
101. Rassoulzadegan M, Grandjean V, Gounon P, Cuzin F. Inheritance of an epigenetic change in the mouse: a new role for RNA. Biochem Soc Trans. 2007;35(Pt 3):623–5. doi:BST0350623 [pii]10.1042/BST0350623.
102. Grandjean V, Gounon P, Wagner N, Martin L, Wagner KD, Bernex F et al. The miR-124-Sox9 paramutation: RNA-mediated epigenetic control of embryonic and adult growth. Development. 2009;136(21):3647–55. doi:136/21/3647 [pii]10.1242/dev.041061.
103. Cenci C, Barzotti R, Galeano F, Corbelli S, Rota R, Massimi L et al. Down-regulation of RNA editing in pediatric astrocytomas: ADAR2 editing activity inhibits cell migration and proliferation. The Journal of biological chemistry. 2008;283(11):7251–60. doi:M708316200 [pii]10.1074/jbc.M708316200.
104. Paz N, Levanon EY, Amariglio N, Heimberger AB, Ram Z, Constantini S et al. Altered adenosine-to-inosine RNA editing in human cancer. Genome Res. 2007;17(11):1586–95. doi:gr.6493107 [pii]10.1101/gr.6493107.
105. Sebastiani P, Montano M, Puca A, Solovieff N, Kojima T, Wang MC, et al. RNA editing genes associated with extreme old age in humans and with lifespan in C. elegans. PLoS ONE. 2009;4(12):e8210. doi:10.1371/journal.pone.0008210.
106. Sodhi MS, Burnet PW, Makoff AJ, Kerwin RW, Harrison PJ. RNA editing of the 5-HT(2C) receptor is reduced in schizophrenia. Mol Psychiatry. 2001;6(4):373–9. doi:10.1038/sj.mp.4000920.
107. Gurevich I, Tamir H, Arango V, Dwork AJ, Mann JJ, Schmauss C. Altered editing of serotonin 2C receptor pre-mRNA in the prefrontal cortex of depressed suicide victims. Neuron. 2002;34(3):349–56. doi:S0896627302006608 [pii].
108. Strahl BD, Allis CD. The language of covalent histone modifications. Nature. 2000;403(6765):41–5. doi:10.1038/47412.

109. Greer EL, Shi Y. Histone methylation: a dynamic mark in health, disease and inheritance. Nat Rev. 2012;13(5):343–57. doi:10.1038/nrg3173.
110. Jenuwein T, Allis CD. Translating the histone code. Science. 2001;293(5532):1074–80.
111. McNairn AJ, Gilbert DM. Epigenomic replication: linking epigenetics to DNA replication. BioEssays. 2003;25(7):647–56. doi:10.1002/bies.10305.
112. Ng HH, Robert F, Young RA, Struhl K. Targeted recruitment of Set1 histone methylase by elongating Pol II provides a localized mark and memory of recent transcriptional activity. Mol Cell. 2003;11(3):709–19.
113. Wang Y, Fischle W, Cheung W, Jacobs S, Khorasanizadeh S, Allis CD. Beyond the double helix: writing and reading the histone code. Novartis Found Symp. 2004;259:3–17; discussion -21, 163–9.
114. Santos-Rosa H, Schneider R, Bannister AJ, Sherriff J, Bernstein BE, Emre NC, et al. Active genes are tri-methylated at K4 of histone H3. Nature. 2002;419(6905):407–11. doi:10.1038/nature01080.
115. Bauer UM, Daujat S, Nielsen SJ, Nightingale K, Kouzarides T. Methylation at arginine 17 of histone H3 is linked to gene activation. EMBO Rep. 2002;3(1):39–44. doi:10.1093/embo-reports/kvf013.
116. Lachner M, O'Carroll D, Rea S, Mechtler K, Jenuwein T. Methylation of histone H3 lysine 9 creates a binding site for HP1 proteins. Nature. 2001;410(6824):116–20. doi:10.1038/35065132.
117. Quina AS, Buschbeck M, Di Croce L. Chromatin structure and epigenetics. Biochem Pharmacol. 2006;72(11):1563–9. doi:10.1016/j.bcp.2006.06.016.
118. Clapier CR, Cairns BR. The biology of chromatin remodeling complexes. Annu Rev Biochem. 2009;78:273–304. doi:10.1146/annurev.biochem.77.062706.153223.
119. Kokavec J, Podskocova J, Zavadil J, Stopka T. Chromatin remodeling and SWI/SNF2 factors in human disease. Frontiers Biosci J Virtual Libr. 2008;13:6126–34.
120. Ko M, Sohn DH, Chung H, Seong RH. Chromatin remodeling, development and disease. Mutat Res. 2008;647(1–2):59–67. doi:10.1016/j.mrfmmm.2008.08.004.
121. Segal E, Widom J. What controls nucleosome positions? Trends Genetics TIG. 2009;25(8):335–43. doi:10.1016/j.tig.2009.06.002.
122. Chi P, Allis CD, Wang GG. Covalent histone modifications–miswritten, misinterpreted and mis-erased in human cancers. Nat Rev Cancer. 2010;10(7):457–69. doi:10.1038/nrc2876.
123. Kurdistani SK. Histone modifications in cancer biology and prognosis. Prog Drug Res. 2011;67:91–106.
124. Urdinguio RG, Sanchez-Mut JV, Esteller M. Epigenetic mechanisms in neurological diseases: genes, syndromes, and therapies. Lancet Neurol. 2009;8(11):1056–72. doi:10.1016/S1474-4422(09)70262-5.
125. Martin DM. Chromatin remodeling in development and disease: focus on CHD7. PLoS Genet. 2010;6(7):e1001010. doi:10.1371/journal.pgen.1001010.
126. Schaefer A, Tarakhovsky A, Greengard P. Epigenetic mechanisms of mental retardation. Prog Drug Res. 2011;67:125–46.
127. Aguzzi A, Rajendran L. The transcellular spread of cytosolic amyloids, prions, and prionoids. Neuron. 2009;64(6):783–90. doi:10.1016/j.neuron.2009.12.016.
128. Goedert M, Clavaguera F, Tolnay M. The propagation of prion-like protein inclusions in neurodegenerative diseases. Trends Neurosci. 2010;33(7):317–25. doi:S0166-2236(10)00055-X [pii]10.1016/j.tins.2010.04.003.
129. Tuite MF, Serio TR. The prion hypothesis: from biological anomaly to basic regulatory mechanism. Nat Rev Mol Cell Biol. 2010;11(12):823–33. doi:nrm3007 [pii]10.1038/nrm3007.
130. Bailey CH, Kandel ER, Si K. The persistence of long-term memory: a molecular approach to self-sustaining changes in learning-induced synaptic growth. Neuron. 2004;44(1):49–57. doi:10.1016/j.neuron.2004.09.017.
131. Si K, Choi YB, White-Grindley E, Majumdar A, Kandel ER. Aplysia CPEB can form prion-like multimers in sensory neurons that contribute to long-term facilitation. Cell. 2010;140(3):421–35. doi:10.1016/j.cell.2010.01.008.

132. Halfmann R, Lindquist S. Epigenetics in the extreme: prions and the inheritance of environmentally acquired traits. Science. 2010;330(6004):629–32. doi:330/6004/629 [pii]10.1126/science.1191081.

133. Halfmann R, Alberti S, Lindquist S. Prions, protein homeostasis, and phenotypic diversity. Trends Cell Biol. 2010;20(3):125–33. doi:S0962-8924(09)00298-0 [pii]10.1016/j.tcb.2009.12.003.

134. Ito K, Caramori G, Lim S, Oates T, Chung KF, Barnes PJ, et al. Expression and activity of histone deacetylases in human asthmatic airways. Am J Respir Crit Care Med. 2002;166(3):392–6.

135. Cosio BG, Mann B, Ito K, Jazrawi E, Barnes PJ, Chung KF, et al. Histone acetylase and deacetylase activity in alveolar macrophages and blood mononocytes in asthma. Am J Respir Crit Care Med. 2004;170(2):141–7. doi:10.1164/rccm.200305-659OC.

136. Su RC, Becker AB, Kozyrskyj AL, Hayglass KT. Altered epigenetic regulation and increasing severity of bronchial hyperresponsiveness in atopic asthmatic children. J Allergy Clin Immunol. 2009;124(5):1116–8. doi:10.1016/j.jaci.2009.08.033.

137. Cosio BG, Tsaprouni L, Ito K, Jazrawi E, Adcock IM, Barnes PJ. Theophylline restores histone deacetylase activity and steroid responses in COPD macrophages. J Exp Med. 2004;200(5):689–95. doi:10.1084/jem.20040416.

138. Ito K, Lim S, Caramori G, Cosio B, Chung KF, Adcock IM, et al. A molecular mechanism of action of theophylline: Induction of histone deacetylase activity to decrease inflammatory gene expression. Proc Natl Acad Sci U S A. 2002;99(13):8921–6. doi:10.1073/pnas.132556899.

139. To Y, Ito K, Kizawa Y, Failla M, Ito M, Kusama T, et al. Targeting phosphoinositide-3-kinase-delta with theophylline reverses corticosteroid insensitivity in chronic obstructive pulmonary disease. Am J Respir Crit Care Med. 2010;182(7):897–904. doi:10.1164/rccm.200906-0937OC.

140. Barnes PJ. Targeting the epigenome in the treatment of asthma and chronic obstructive pulmonary disease. Proc Am Thorac Soc. 2009;6(8):693–6. doi:10.1513/pats.200907-071DP.

141. Jeevan-Raj BP, Robert I, Heyer V, Page A, Wang JH, Cammas F, et al. Epigenetic tethering of AID to the donor switch region during immunoglobulin class switch recombination. The J Exp Med. 2011;. doi:10.1084/jem.20110118.

142. Stanlie A, Aida M, Muramatsu M, Honjo T, Begum NA. Histone3 lysine4 trimethylation regulated by the facilitates chromatin transcription complex is critical for DNA cleavage in class switch recombination. Proc Natl Acad Sci U S A. 2010. doi:1016923108 [pii]10.1073/pnas.1016923108.

143. Daniel JA, Santos MA, Wang Z, Zang C, Schwab KR, Jankovic M et al. PTIP promotes chromatin changes critical for immunoglobulin class switch recombination. Science. 2010. doi:science.1187942 [pii]10.1126/science.1187942.

144. Kuang FL, Luo Z, Scharff MD. H3 trimethyl K9 and H3 acetyl K9 chromatin modifications are associated with class switch recombination. Proc Natl Acad Sci U S A. 2009;106(13):5288–93. doi:0901368106 [pii]10.1073/pnas.0901368106.

145. Erazo A, Kutchukhidze N, Leung M, Christ AP, Urban JF Jr, de Curotto LMA, et al. Unique maturation program of the IgE response in vivo. Immunity. 2007;26(2):191–203.

146. Xiong H, Dolpady J, Wabl M, de Curotto LMA, Lafaille JJ. Sequential class switching is required for the generation of high affinity IgE antibodies. J Exp Med. 2012;. doi:10.1084/jem.20111941.

147. Grausenburger R, Bilic I, Boucheron N, Zupkovitz G, El-Housseiny L, Tschismarov R, et al. Conditional deletion of histone deacetylase 1 in T cells leads to enhanced airway inflammation and increased Th2 cytokine production. J Immunol. 2010;185(6):3489–97. doi:10.4049/jimmunol.0903610.

148. Vercelli D. Genetics, epigenetics, and the environment: switching, buffering, releasing. The J Allergy Clin Immunol. 2004;113(3):381–6; quiz 7. doi:10.1016/j.jaci.2004.01.752.

149. Pascual M, Davila I, Isidoro-Garcia M, Lorente F. Epigenetic aspects of the allergic diseases. Front Biosci (Schol Ed). 2010;2:815–24.

150. Vercelli D. Discovering susceptibility genes for asthma and allergy. Nat Rev Immunol. 2008;8(18301422):169–82.
151. Lee JS, Smith E, Shilatifard A. The language of histone crosstalk. Cell. 2010;142(5):682–5. doi:10.1016/j.cell.2010.08.011.
152. Hebbes TR, Thorne AW, Crane-Robinson C. A direct link between core histone acetylation and transcriptionally active chromatin. EMBO J. 1988;7(5):1395–402.
153. Gilmour DS, Lis JT. Detecting protein-DNA interactions in vivo: distribution of RNA polymerase on specific bacterial genes. Proc Natl Acad Sci U S A. 1984;81(14):4275–9.
154. Solomon MJ, Larsen PL, Varshavsky A. Mapping protein-DNA interactions in vivo with formaldehyde: evidence that histone H4 is retained on a highly transcribed gene. Cell. 1988;53(6):937–47.
155. Gilmour DS, Lis JT. In vivo interactions of RNA polymerase II with genes of Drosophila melanogaster. Mol Cell Biol. 1985;5(8):2009–18.
156. Aparicio O, Geisberg JV, Sekinger E, Yang A, Moqtaderi Z, Struhl K. Chromatin immunoprecipitation for determining the association of proteins with specific genomic sequences in vivo. In: Frederick MA et al. , editors. Current protocols in molecular biology. 2005;Chapter 21:Unit 21 3. doi:10.1002/0471142727.mb2103s69.
157. Kim TH, Barrera LO, Ren B. ChIP-chip for genome-wide analysis of protein binding in mammalian cells. In: Frederick MA et al. , editors. Current protocols in molecular biology. 2007;Chapter 21:Unit 21 13. doi:10.1002/0471142727.mb2113s79.
158. Moqtaderi Z, Struhl K. Defining in vivo targets of nuclear proteins by chromatin immunoprecipitation and microarray analysis. In: Frederick MA et al. ,editors. Current protocols in molecular biology. 2004;Chapter 21:Unit 21 9. doi:10.1002/0471142727.mb2109s68.
159. Raha D, Hong M, Snyder M. ChIP-Seq: a method for global identification of regulatory elements in the genome. In: Frederick MA et al. ,editors. Current protocols in molecular biology. 2010;Chapter 21:Unit 21 19 1–4. doi:10.1002/0471142727.mb2119s91.
160. Ren B, Robert F, Wyrick JJ, Aparicio O, Jennings EG, Simon I et al. Genome-wide location and function of DNA binding proteins. Science. 2000;290(5500):2306–9. doi:10.1126/science.290.5500.2306.
161. Liu CL, Schreiber SL, Bernstein BE. Development and validation of a T7 based linear amplification for genomic DNA. BMC Genomics. 2003;4(1):19. doi:10.1186/1471-2164-4-19.
162. Mikkelsen TS, Ku M, Jaffe DB, Issac B, Lieberman E, Giannoukos G, et al. Genome-wide maps of chromatin state in pluripotent and lineage-committed cells. Nature. 2007;448(7153):553–60. doi:10.1038/nature06008.
163. Barski A, Cuddapah S, Cui K, Roh TY, Schones DE, Wang Z, et al. High-resolution profiling of histone methylations in the human genome. Cell. 2007;129(4):823–37. doi:10.1016/j.cell.2007.05.009.
164. Robertson AG, Bilenky M, Tam A, Zhao Y, Zeng T, Thiessen N, et al. Genome-wide relationship between histone H3 lysine 4 mono- and tri-methylation and transcription factor binding. Genome Res. 2008;18(12):1906–17. doi:10.1101/gr.078519.108.
165. Robertson G, Hirst M, Bainbridge M, Bilenky M, Zhao Y, Zeng T, et al. Genome-wide profiles of STAT1 DNA association using chromatin immunoprecipitation and massively parallel sequencing. Nat Methods. 2007;4(8):651–7. doi:10.1038/nmeth1068.
166. Johnson DS, Mortazavi A, Myers RM, Wold B. Genome-wide mapping of in vivo protein-DNA interactions. Science . 2007;316(5830):1497–502. doi:10.1126/science.1141319.
167. Laird PW. Principles and challenges of genomewide DNA methylation analysis. Nat Rev. 2010;11(3):191–203. doi:10.1038/nrg2732.
168. Martin-Subero JI, Esteller M. Profiling epigenetic alterations in disease. Adv Exp Med Biol. 2011;711:162–77.
169. Hayatsu H. Bisulfite modification of nucleic acids and their constituents. Prog Nucleic Acid Res Mol Biol. 1976;16:75–124.
170. Frommer M, McDonald LE, Millar DS, Collis CM, Watt F, Grigg GW, et al. A genomic sequencing protocol that yields a positive display of 5-methylcytosine residues in individual DNA strands. Proc Natl Acad Sci U S A. 1992;89(5):1827–31.

171. Colella S, Shen L, Baggerly KA, Issa JP, Krahe R. Sensitive and quantitative universal pyrosequencing methylation analysis of CpG sites. Biotechniques. 2003;35(1):146–50.
172. Colyer HA, Armstrong RN, Sharpe DJ, Mills KI. Detection and analysis of DNA methylation by pyrosequencing. Methods Mol Biol. 2012;863:281–92. doi:10.1007/978-1-61779-612-8_17.
173. Tost J, Gut IG. DNA methylation analysis by pyrosequencing. Nat Protoc. 2007;2(9):2265–75. doi:10.1038/nprot.2007.314.
174. Tost J, El abdalaoui H, Gut IG. Serial pyrosequencing for quantitative DNA methylation analysis. Biotechniques. 2006;40(6):721–2, 4, 6.
175. Tost J, Gut IG. Analysis of gene-specific DNA methylation patterns by pyrosequencing technology. Methods Mol Biol. 2007;373:89–102. doi:10.1385/1-59745-377-3:89.
176. Herman JG, Graff JR, Myohanen S, Nelkin BD, Baylin SB. Methylation-specific PCR: a novel PCR assay for methylation status of CpG islands. Proc Natl Acad Sci U S A. 1996;93(18):9821–6.
177. Xiong Z, Laird PW. COBRA: a sensitive and quantitative DNA methylation assay. Nucleic Acids Res. 1997;25(12):2532–4.
178. Ehrich M, Nelson MR, Stanssens P, Zabeau M, Liloglou T, Xinarianos G, et al. Quantitative high-throughput analysis of DNA methylation patterns by base-specific cleavage and mass spectrometry. Proc Natl Acad Sci U S A. 2005;102(44):15785–90. doi:10.1073/pnas.0507816102.
179. Huang TH, Laux DE, Hamlin BC, Tran P, Tran H, Lubahn DB. Identification of DNA methylation markers for human breast carcinomas using the methylation-sensitive restriction fingerprinting technique. Cancer Res. 1997;57(6):1030–4.
180. Yan PS, Potter D, Deatherage DE, Huang TH, Lin S. Differential methylation hybridization: profiling DNA methylation with a high-density CpG island microarray. Methods Mol Biol. 2009;507:89–106. doi:10.1007/978-1-59745-522-0_8.
181. Khulan B, Thompson RF, Ye K, Fazzari MJ, Suzuki M, Stasiek E et al. Comparative isoschizomer profiling of cytosine methylation: the HELP assay. Genome Res. 2006;16(8):1046–55. doi:gr.5273806 [pii]10.1101/gr.5273806.
182. Thompson RF, Reimers M, Khulan B, Gissot M, Richmond TA, Chen Q et al. An analytical pipeline for genomic representations used for cytosine methylation studies. Bioinformatics. 2008;24(9):1161–7. doi:btn096 [pii]10.1093/bioinformatics/btn096.
183. Suzuki M, Jing Q, Lia D, Pascual M, McLellan A, Greally JM. Optimized design and data analysis of tag-based cytosine methylation assays. Genome Biol. 2010;11(4):R36. doi:gb-2010-11-4-r36 [pii]10.1186/gb-2010-11-4-r36.
184. Jing Q, McLellan A, Greally JM, Suzuki M. Automated computational analysis of genome-wide DNA methylation profiling data from HELP-tagging assays. Methods Mol Biol. 2012;815:79–87. doi:10.1007/978-1-61779-424-7_7.
185. Weber M, Hellmann I, Stadler MB, Ramos L, Paabo S, Rebhan M, et al. Distribution, silencing potential and evolutionary impact of promoter DNA methylation in the human genome. Nat Genet. 2007;39(4):457–66. doi:10.1038/ng1990.
186. Meissner A, Gnirke A, Bell GW, Ramsahoye B, Lander ES, Jaenisch R. Reduced representation bisulfite sequencing for comparative high-resolution DNA methylation analysis. Nucleic Acids Res. 2005;33(18):5868–77. doi:10.1093/nar/gki901.
187. Gu H, Smith ZD, Bock C, Boyle P, Gnirke A, Meissner A. Preparation of reduced representation bisulfite sequencing libraries for genome-scale DNA methylation profiling. Nat Protoc. 2011;6(4):468–81. doi:10.1038/nprot.2010.190.
188. Sun Z, Baheti S, Middha S, Kanwar R, Zhang Y, Li X, et al. SAAP-RRBS: streamlined analysis and annotation pipeline for reduced representation bisulfite sequencing. Bioinformatics. 2012;. doi:10.1093/bioinformatics/bts337.

Chapter 4
Epigenetics, Environment, and Allergic Diseases

In 1989, David Strachan, an English epidemiologist, proposed a theory, colloquially known as the 'hygiene hypothesis', offering a possible explanation for the rising prevalence of allergic diseases [1]. The hypothesis postulated that the family size, position within the family, or infections suffered during childhood most likely transmitted by unhygienic contact with older siblings, could play an important protective role in the increased prevalence of allergic diseases. The idea was met with skepticism at first, because the dominant immune thinking favored the idea that infections could predispose the individual to suffer from allergic diseases [2]. However, in the early 1990s, two different populations of T helper lymphocytes were described, named Th1 and Th2 [3]. This discovery changed the previous perspective, as it was found that in laboratory animals the induction of a Th1 response against viral or bacterial infections resulted in detriment of Th2 responses, which are involved in allergic responses. Then the hygiene hypothesis started again to be reconsidered, and the paradigm has continued until nowadays. It suggests that the exposure to infectious agents during early childhood can induce Th1 responses, while it has the potential to suppress the development of Th2-induced diseases later in life.

The molecular mechanisms underlying the predisposition to suffer allergic diseases are still being explored [4–7]. The rapid and recent increase in the prevalence of allergic diseases in Western countries could reflect the recent changes in the environment and in the lifestyle, which would affect more aggressively to individuals genetically predisposed to suffer allergic diseases. While challenging, many studies have attempted to delve into gene–environment interactions that promote the development, increase the severity, or limit a positive development of allergic inflammation [8–11]. There are now evidences that exposure to a single microbial product can exert even opposite effects on individuals prone to develop an allergic response, depending on their allergic phenotype, adding extra levels of complexity to the pathogenesis of allergy.

M. Pascual and S. Roa, *Epigenetic Approaches to Allergy Research*,
SpringerBriefs in Genetics, DOI: 10.1007/978-1-4614-6366-5_4,
© The Author(s) 2013

4.1 *In Utero* and Early Life: Epigenetics and Environmental Aggressions

4.1.1 Origins of Health and Illness During Development (Developmental Origins of Health and Disease)

In 1992, a new hypothesis, known as "Baker Hypothesis", "Early Origins of Adult Diseases", or "Developmental Origins of Health and Disease" was proposed [12, 13]. This hypothesis suggested that different environmental exposures during the intrauterine period could alter the genetic (or epigenetic) program affecting the development of certain organs, and thus, determining different physiological and metabolic conditions later in life. This implies that the combined effect of environmental and genetic factors may have an important role in predisposing an individual to develop a disease later in life (see Fig. 4.1).

Epidemiological studies have shown that certain environmental factors that occur *in Utero* or early in life may influence the subsequent development of certain chronic diseases such as cancer, cardiovascular disease, diabetes, obesity, or behavioral disorders as schizophrenia [14–17]. One of the general mechanisms by which prenatal or postnatal environmental exposure could alter the phenotype during later stages of life could be explained through epigenetic mechanisms. The fact that the aberrations in the epigenome of somatic cells can be transmitted during mitotic cell divisions provides a possible mechanism by which environmental effects on the epigenome may have implications for long-term gene expression [18, 19].

In this sense, studies in animal models show that different environmental stimuli that occur during the early stages of life, like addition of certain nutritional

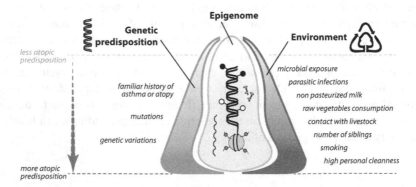

Fig. 4.1 Impact of the genetic and environmental elements in the epigenome. This gradated pyramid illustrates how the increased number of genetic variations or inherited genetic predisposition at specific genes related to the allergic reaction, together with exposures to environmental situations that can have an impact in the immune response, may cooperate directly or indirectly with the plasticity of the epigenome (DNA modifications, regulatory RNAs, chromatin remodeling, or histone modifications) to increase the risk of atopic sensitization

supplements [20, 21], xenobiotics [22], behavioral problems [23], small doses of radiation [24], and smoke or snuff [25], can have a tremendous impact on epigenetic programming, and therefore on the subsequent development of disease.

4.2 Epigenetics and Allergic Diseases

Allergic responses are characterized by a systemic inflammation that results from a disproportionate Th2 response, which results in an increased expression of IL-4, IL-5, IL-9, and IL-13, as well as an increase in plasma levels of IgE. Genetically, allergic diseases are defined as (i) *polygenic*, because a large number of genes have been described to have pivotal roles in the predisposition, onset, and progression of the disease; (ii) *genetically heterogeneous*, because different combinations of candidate genes are associated with the development of allergic diseases; and, finally, (iii) *pleiotropic*, because a single gene can be involved in more than one phenotypic trait.

The introduction of epigenetics in the context of allergic diseases has opened a new field of study that may hold the key to understand unresolved issues such as the phenotypic differences between monozygotic twins, age of establishment of the disease and its severity, the increased incidence of these diseases in women compared to men, or the fact that if the mother is allergic, the possibilities for the offspring to develop any allergic disease are greater than in case is the father the one who has the allergic disease [26, 27].

It has been proposed that in order to consider epigenetic modifications as real candidates to influence the etiology of complex diseases, it should meet at least these three characteristics [28, 29]:

1. *It must influence the phenotype*: monozygotic twins share the same genetic information, but show a discrepancy in the prevalence of allergic diseases reaching 25 % of cases. In this sense, epigenetic modifications could explain also differences in prevalence between sexes, age of disease onset, or its severity [8].
2. *It has to be heritable*: it has been shown that epigenetic modifications are transmitted to offspring. Moreover, *in vitro* culture of T cells has shown that the epigenetic patterns are maintained for 20 divisions in the absence of any external stimulus, proving the heritability of epigenetic patterns in somatic cells. This is particularly important when considering issues such as immunological memory [30].
3. *It has to respond dynamically to environmental variables*: it is known that changes in epigenetic patterns occur more frequently than changes that may occur in the sequence of base pairs of DNA [28].

In recent years, various studies have attempted to elucidate the possible connection between various environmental exposures and epigenetics in atopic diseases. We will review some of the environmental factors that have been proposed to influence the predisposition to allergic diseases, such as diet, exposure to polycyclic aromatic hydrocarbons, the rural environment, or smoking.

4.2.1 Diet

Modern diet, present in most westernized countries, is far away different from the diet of previous generations in those same countries. Our contemporary diet is now frequently based on foods that have been processed, modified, stored, and sometimes transported long distances. In addition, our diet also includes fewer vegetables and their nutritional content has probably also decreased. This trend has led to the hypothesis that these changes in diet may be related to the increased prevalence of allergic diseases [31].

Hollingsworth et al. [32] based their work on the hypothesis that certain changes to the diet could possibly have an effect on the subsequent development of allergic diseases. It had been shown that adequate intake of folic acid is essential for proper formation of the baby's neural tube, which also prevents possible birth defects [33, 34]. This dietary supplement is a donor source of methyl groups, and has been shown that, at the same time, can cause changes in DNA methylation and thus affect the normal gene expression patterns [35]. Because in Western countries many pregnant women are advised to supplement their diet with folic acid, the authors of this study raised the possibility that besides protecting the fetus from future complications, it could also increase the chances of developing asthma in the future.

To prove their hypothesis, they examined the global levels of DNA methylation in C57BL/6J mice. They used lung tissue from embryo mice whose mothers had been supplemented with different diets: two of them received a diet rich in folate, whereas the other two were supplemented with low doses of folate. To assess DNA methylation levels, the authors used a technique known as MSDK (for its acronym Methylation Sensitive Digital Karyotyping) and found several genes that showed different levels of DNA methylation, as well as expression between the two groups. Among these genes, the authors reported an increase on methylation and reduced expression of *Runx3*, which others have associated with a skewed Th2 response [36]. Consistently, RUNX3-deficient mice spontaneously develop an asthma-like phenotype [37]. To our knowledge, this is the first study *in vivo* showing that diet can have a direct impact on health, particularly at the onset of asthmatic diseases. The maternal administration of folate-rich diets during pregnancy, which seems to be a period of epigenetic vulnerability, may have an impact on the methylation levels and predisposition to allergic diseases, while the periods of lactation or adulthood appear less significant to the development and severity of respiratory diseases.

Retinoic acid (RA) is known to be an important regulator in the immune system, and an imbalance in the RA metabolism may have an immunological impact, being also related to the susceptibility to allergic diseases [38]. There is an interesting body of work correlating the effects of low retinol concentration during the first months of life and the development of allergy during adulthood [39, 40]. This is supported by *in vitro* studies, where cytokine-stimulated human B cells cultured in presence of ATRA (all-*trans* retinoic acid, the active metabolite of vitamin A) seem to have impaired the production of IgE [41]. However, murine models seem to contradict these theories. In this sense, it has been shown that poor diets in vitamin A impair the development of allergic responses in ovalbumin-challenged mice, whereas high levels of vitamin A intake showed a shift toward Th2 with an aggravation of allergic symptoms [42] and up-regulation of IgE in serum [43].

In a study published in 2011, purified CD19$^+$ B cells from house dust mite-allergic patients were shown to have differential DNA methylation patterns when compared with nonallergic subjects, including healthy donors and nonatopic asthmatic patients [44]. The authors showed differences in methylation of genes involved in processes critical for immune responses, as well as genes involved in RA signaling pathways in these dust allergic patients. Interestingly, the authors focused on the *CYP26A1* gene, which is involved in retinoic acid catabolism and was just associated with allergic diseases in this study for the first time.

4.2.2 Exposure to Polycyclic Aromatic Hydrocarbons

Recently, it has been hypothesized that *in utero* exposure to polycyclic aromatic hydrocarbons (PAH) coming from urban traffic, could be a risk factor and help to explain the increase of asthmatic diseases in young children [45, 46]. Perera et al. studied Hispanic and African American nonsmoking mothers who were monitored regularly during pregnancy in an attempt to explore the epigenetic impact of transplacental exposures to PAHs. After isolation of T cells from umbilical cord, the group focused on the global methylation patterns through MSRF (for its acronym in *M*ethylation *S*ensitive *R*estriction *F*ingerprint) [47, 48]. The authors reported that changes in methylation of the promoter region of *ASCL3* gene (acyl-CoA synthetase long-chain family member 3), were significantly associated with maternal exposure to PAHs, as well as appearance of asthma symptoms during the first 5 years of life. The study, which started with a preliminary group of 20 subjects was expanded to 56 placental tissues, in order to confirm the initial results on the candidate gene *ASCL3*. The information on the levels of methylation, obtained by MSPCR (for its acronym *M*ethyl *S*ensitive *PCR*) showed a positive association between the level of exposure to these compounds and methylation. Although further studies are required to better understand the impact of PAH compounds during pregnancy and childhood, these results support the emerging theory that early live exposures may have an effect on the predisposition and development of later life diseases.

4.2.3 Rural Environment

Some epidemiological studies support the idea of a protective role of the rural environment during childhood against allergic diseases [49, 50]. The hypothesis lies on the fact that during early childhood, a life in contact with stables and barns, with domestic animals, and consuming unpasteurized milk or row vegetables enables a direct contact with a variety of microbial agents. These exposures during early life would enhance Th1 responses over Th2 responses, a bias that might confer some protection against the development of allergic diseases later in life.

Shaub et al. [51] set out to determine whether this protective effect that seems to take place in rural environments exerts it effects during *in utero* life. They selected 82 healthy mothers (22 living in rural and 60 of them living in urban environments), with no complications during pregnancy, and isolated mononuclear cells from the umbilical cord. Of the total cell counts, the authors focused on regulatory T cells (Treg), due to the important role that these cells exert during maturation and polarization of the different T cell populations. Upon stimulation of cells with different antigen combinations (peptidoglycan, PHA, and *Dermatophagoides pteronyssinus*) for 3 consecutive days, the authors analyzed the methylation of FOXP3 by RT-PCR, and found no significant differences in methylation levels between groups. However, they reported a higher number of Treg cells associated with a decrease in Th2 response in samples collected from mothers living in the rural environment, when compared with the same cell population obtained from mothers that lived in urban areas.

4.2.4 Smoking

It is well known that continuous exposure to environmental tobacco smoke is a major cause of irreversible chronic inflammation in the lung, thus leading to worsening of the respiratory capacity and chronic obstructive pulmonary disease (COPD) [52]. It is well known that the macrophage count of these patients increases concomitantly with the severity of the disease. Moreover, the proinflammatory cytokines secreted by these cells are also increased in cases where the patient is or has been a smoker [53]. In terms of epigenetic modifications, it is known that the activity of histone deacetylases (HDAC1, HDAC2, and HDAC3) in macrophages are decreased after treatment with smoke extracts, as well as its cellular concentration [54]. Another similar observation was made in rat lungs that exerted an acute inflammatory response after tobacco smoke exposure [55, 56]. Taken together these observations, it is not surprising that due to the great importance of these enzymes in epigenetic balance, any disruption in their activities may have an impact on the expression of some of the most important proinflammatory cytokines and contribute to the pathogenesis of the disease.

In murine models, both active and passive smoking have been shown to predispose to allergic sensitization. A possible mechanism proposed to explain this observation, as has been proposed previously for other harmful substances [57], is that the tobacco smoke could increase antigen presentation either by mechanisms of adsorption to the allergen or by structural changes in the allergen itself [58].

In addition, some studies have shown that exposure to tobacco smoke may alter gene expression, promoting hypermethylation of certain gene promoters, which are normally expressed in lungs of healthy individuals [59, 60]. Moreover, the familiar history of smoking habits can also influence the subsequent development of asthma for future generations and not only if the mother has been an active smoker during pregnancy, but also if the grandmother has been an active smoker during pregnancy of the mother can determine the chances that the offspring has to develop asthma related diseases [61]. Overall, it is striking to think that by affecting the epigenome, the exposure to smoke may affect susceptibility to allergic diseases not only in the individual, but also in the following generations.

4.3 Epigenetics and Th2-Cell Phenotype in Atopic Disorders

T cells have an important role in orchestrating host defense as well as in regulating the immune response. Several subpopulations of T cells have been described, each of them showing different functional properties and epigenetic marks [62]. Cell lineage decisions in T-helper cells are influenced by the surrounding cytokine milieu at the site of antigen encounter [63, 64]. T helper type 1 (Th1) cells are involved in the host defense against intracellular infections through the production of almost exclusively IFN-gamma [62]. On the other hand, Th2 cells produce IL-4, IL-5, and IL-13 and are involved in the defense against helminthes. In the case of allergic reactions, the balance Th1/Th2 is skewed to an exaggerated Th2 response against substances that are harmless under normal conditions [65]. The recently described Th17 cells are involved in neutrophil-mediated protection against extracellular bacteria and fungi, by producing cytokines IL-17A, IL-17F, IL-21, IL-22, and IL-26 [66]. Regulatory T cells (Treg) are characterized by a continuous expression of FOXP3, and have a crucial role on maintaining homeostasis of the immune system and in preventing the autoimmune reactivity of self-reactive T cells [67, 68].

Th1 and Th2 development are mutually antagonistic processes and, as already mentioned, in allergic diseases the Th1/Th2 balance is off, presenting a skewed Th2 response and IL-4, IL-5, and IL-13 release against substances that are harmless under normal circumstances. Interestingly, in the human genome, the genes encoding interleukins 4, 5, and 13 are in the same chromosomal region (5q31.1) [65]. The expression of these cytokines is accompanied by two events: (1) an increased transcription of the GATA3 transcription factor, and (2) a process of chromatin remodeling that promotes a physiological 'open' state that allows gene transcription [69]. GATA3 expression is essential for the differentiation toward a Th2 lineage being indispensable for the production of IL-5 and IL-13, but not IL-4 [70]. The binding sites of GATA3 are found in the promoter region of IL-5 and IL-13, indicating the crucial role that this transcription factor has in the transcription of these interleukins [71]. Inhibition of GATA3 in Th2 responses implies an inhibition in the expression of IL-3 and IL-5, and thus results in the decline of the Th2 lineage specification [70].

Chromatin remodeling processes were initially demonstrated by digestion with DNase I. In transcriptionally silenced loci, the DNA is highly condensed around histones, forming a superstructure known as heterochromatin. In this conformation, the DNA is resistant to digestion with DNase I. However, when the DNA is in a transcriptional active state (known as euchromatin), it becomes susceptible to this digestion and is defined as a hypersensitive region (HS). Differentiation of T cells into mature Th2 cells seems to be associated with the appearance of HS sites within the IL-4/IL-13 cytokine gene cluster. Several loci were defined in the 5q31.1 region based on their susceptibility to the DNase I digestion [65]. Among these loci, some correspond to the conserved noncoding sequence 1 (NCS1) located in an intergenic region between IL-4 and IL-13 [72]. The hypersensitive regions HS-II in the second intron of IL-4 and HS-V at the promoter region could also be characterized by this approach, and defined an essential enhancer with potential to regulate IL-4 gene expression [73]. On the other hand, the HS-IV region and a permissive chromatin structure in all helper T cells, associated with IL-4 silencing during Th1 differentiation, suggest a mechanism of IL-4 silencing in Th1-mediated immunity that differs from Th2 responses [74]. Of particular importance is the locus RHS7, which is accessible to digestion with DNase I at the beginning of the Th2 differentiation process [75].

Coming back to the discussion about transcription factors being recruited to the cytokine gene cluster region, several studies have postulated that GATA-3 could be implicated in the remodeling of chromatin within the IL-4/IL-13 locus, modulating the appearance of DNase I sensitive sites and promoting Th2 differentiation [76, 77]. Furthermore, trimethylation of lysine 4 at histone 3 (H3K4me3), which is characteristic of transcriptionally active sites, is present in the IL-4/IL-13 locus of Th2 cells [78, 79]. On the other hand, trimethylation of lysine 27 at histone 3 (H3K27me3), which is characteristic of transcriptionally silenced regions, is present in the same locus in Th1 cells [80]. In addition to these changes in the conformation of chromatin, changes in DNA methylation around the IL-4 locus have been also described. In this sense, it has been shown that this locus presents a gain of methylation (repressive mark) in Th1 cells, which appears to be lost in Th2 cells [81].

All these results illustrate the importance of chromatin accessibility, and suggest the existence of epigenetic mechanisms of regulation of the transcriptional accessibility of specific genomic areas during the differentiation of Th1 and Th2 cells.

References

1. Strachan DP. Hay fever, hygiene, and household size. BMJ. 1989;299(6710):1259–60.
2. Busse WW. The relationship between viral infections and onset of allergic diseases and asthma. Clin Exp Allergy. 1989;19(1):1–9.

3. Romagnani S. Type 1 T helper and type 2 T helper cells: functions, regulation and role in protection and disease. Int J Clin Lab Res. 1991;21(2):152–8.
4. Yazdanbakhsh M, Rodrigues LC. Allergy and the hygiene hypothesis: the Th1/Th2 counterregulation can not provide an explanation. Wien Klin Wochenschr. 2001;113(23–24):899–902.
5. Yazdanbakhsh M, Kremsner PG, van Ree R. Allergy, parasites, and the hygiene hypothesis. Science (New York). 2002;296(5567):490–4. doi:10.1126/science.296.5567.490.
6. Fallon PG, Mangan NE. Suppression of TH2-type allergic reactions by helminth infection. Nat Rev Immunol. 2007;7(3):220–30. doi:10.1038/nri2039.
7. Romagnani S. Coming back to a missing immune deviation as the main explanatory mechanism for the hygiene hypothesis. J Allergy Clin Immunol. 2007;119(6):1511–3. doi:10.1016/j.jaci.2007.04.005.
8. Vercelli D. Genetics, epigenetics, and the environment: switching, buffering, releasing. The J Allergy Clin Immunol 2004;113(3):381–386; quiz 7. doi:10.1016/j.jaci.2004.01.752.
9. Pascual M, Davila I, Isidoro-Garcia M, Lorente F. Epigenetic aspects of the allergic diseases. Front Biosci (Schol Ed). 2010;2:815–24.
10. Vercelli D. Discovering susceptibility genes for asthma and allergy. Nat Rev Immunol. 2008;8(18301422):169–82.
11. Cookson W. The immunogenetics of asthma and eczema: a new focus on the epithelium. Nat Rev Immunol. 2004;4(12):978–88. doi:10.1038/nri1500.
12. Barker DJ. The origins of the developmental origins theory. J Intern Med. 2007;261(5):412–7. doi:10.1111/j.1365-2796.2007.01809.x.
13. Barker DJ. The developmental origins of adult disease. J Am Coll Nutr. 2004;23(6 Suppl):588S–95S.
14. Gluckman PD, Hanson MA, Cooper C, Thornburg KL. Effect of in utero and early-life conditions on adult health and disease. N Engl J Med. 2008;359(1):61–73. doi:10.1056/NEJMra0708473.
15. Gluckman PD, Hanson MA, Beedle AS. Early life events and their consequences for later disease: a life history and evolutionary perspective. Am J Hum Biol Official J Hum Biol Counc. 2007;19(1):1–19. doi:10.1002/ajhb.20590.
16. St Clair D, Xu M, Wang P, Yu Y, Fang Y, Zhang F, et al. Rates of adult schizophrenia following prenatal exposure to the Chinese famine of 1959–1961. Jama. 2005;294(5):557–62. doi:1 0.1001/jama.294.5.557.
17. van Os J, Selten JP. Prenatal exposure to maternal stress and subsequent schizophrenia. The May 1940 invasion of The Netherlands. British J Psychiatry J Ment Sci. 1998;172:324–6.
18. Dolinoy DC, Jirtle RL. Environmental Epigenomics in Human Health and Disease. Environ Mol Mutagen. 2008;49(1):4–8. doi:10.1002/em.20366.
19. Jirtle RL, Skinner MK. Environmental epigenomics and disease susceptibility. Nat Rev. 2007;8(4):253–62. doi:10.1038/nrg2045.
20. Wolff GL, Kodell RL, Moore SR, Cooney CA. Maternal epigenetics and methyl supplements affect agouti gene expression in Avy/a mice. FASEB J. 1998;12(11):949–57.
21. Waterland RA, Jirtle RL. Transposable elements: targets for early nutritional effects on epigenetic gene regulation. Mol Cell Biol. 2003;23(15):5293–300.
22. Ho SM, Tang WY, Belmonte de Frausto J, Prins GS. Developmental exposure to estradiol and bisphenol A increases susceptibility to prostate carcinogenesis and epigenetically regulates phosphodiesterase type 4 variant 4. Canc Res. 2006;66(11):5624–32. doi:10.1158/0008-5472.CAN-06-0516.
23. Weaver IC, Champagne FA, Brown SE, Dymov S, Sharma S, Meaney MJ, et al. Reversal of maternal programming of stress responses in adult offspring through methyl supplementation: altering epigenetic marking later in life. J Neurosci. 2005;25(47):11045–54. doi:10.152 3/JNEUROSCI.3652-05.2005.
24. Koturbash I, Baker M, Loree J, Kutanzi K, Hudson D, Pogribny I, et al. Epigenetic dysregulation underlies radiation-induced transgenerational genome instability in vivo. Int J Radiat Oncol Biol Phys. 2006;66(2):327–30. doi:10.1016/j.ijrobp.2006.06.012.

25. Breton CV, Byun HM, Wenten M, Pan F, Yang A, Gilliland FD. Prenatal tobacco smoke exposure affects global and gene-specific DNA methylation. Am J Respir Crit Care Med. 2009;180(5):462–7. doi:10.1164/rccm.200901-0135OC.

26. Lovinsky-Desir S, Miller RL. Epigenetics, asthma, and allergic diseases: a review of the latest advancements. Curr Allergy Asthma Rep. 2012;12(3):211–20. doi:10.1007/s11882-012-0257-4.

27. Miller RL, Ho SM. Environmental epigenetics and asthma: current concepts and call for studies. Am J Respir Crit Care Med. 2008;177(6):567–73. doi:10.1164/rccm.200710-1511PP.

28. Hatchwell E, Greally JM. The potential role of epigenomic dysregulation in complex human disease. Trends Genet. 2007;23(11):588–95. doi:10.1016/j.tig.2007.08.010.

29. Ptak C, Petronis A. Epigenetics and complex disease: from etiology to new therapeutics. Annu Rev Pharmacol Toxicol. 2008;48:257–76. doi:10.1146/annurev.pharmtox.48.113006.094731.

30. Richards EJ. Inherited epigenetic variation–revisiting soft inheritance. Nat Rev. 2006;7(5):395–401. doi:10.1038/nrg1834.

31. Devereux G. The increase in the prevalence of asthma and allergy: food for thought. Nat Rev Immunol. 2006;6(11):869–74. doi:10.1038/nri1958.

32. Hollingsworth JW, Maruoka S, Boon K, Garantziotis S, Li Z, Tomfohr J, et al. In utero supplementation with methyl donors enhances allergic airway disease in mice. J Clin Invest. 2008;118(10):3462–9. doi:10.1172/JCI34378.

33. Taparia S, Gelineau-van Waes J, Rosenquist TH, Finnell RH. Importance of folate-homocysteine homeostasis during early embryonic development. Clin Chem Lab Med CCLM/FESCC. 2007;45(12):1717–27. doi:10.1515/CCLM.2007.345.

34. Rosenquist TH, Finnell RH. Genes, folate and homocysteine in embryonic development. Proc Nutr Soc. 2001;60(1):53–61.

35. Feinberg AP, Tycko B. The history of cancer epigenetics. Nat Rev Canc. 2004;4(2):143–53. doi:10.1038/nrc1279.

36. Djuretic IM, Levanon D, Negreanu V, Groner Y, Rao A, Ansel KM. Transcription factors T-bet and Runx3 cooperate to activate Ifng and silence Il4 in T helper type 1 cells. Nat Immunol. 2007;8(2):145–53. doi:10.1038/ni1424.

37. Fainaru O, Shseyov D, Hantisteanu S, Groner Y. Accelerated chemokine receptor 7-mediated dendritic cell migration in Runx3 knockout mice and the spontaneous development of asthma-like disease. Proc Natl Acad Sci USA. 2005;102(30):10598–603. doi:10.1073/pnas.0504787102.

38. Iwata M, Eshima Y, Kagechika H. Retinoic acids exert direct effects on T cells to suppress Th1 development and enhance Th2 development via retinoic acid receptors. Int Immunol. 2003;15(8):1017–25.

39. Pesonen M, Kallio MJ, Siimes MA, Ranki A. Retinol concentrations after birth are inversely associated with atopic manifestations in children and young adults. Clin Exp Allergy. 2007;37(1):54–61. doi:10.1111/j.1365-2222.2006.02630.x.

40. Al Senaidy AM. Serum vitamin A and beta-carotene levels in children with asthma. J Asthma Official J Assoc Care Asthma. 2009;46(7):699–702. doi:10.1080/02770900903056195.

41. Scheffel F, Heine G, Henz BM, Worm M. Retinoic acid inhibits CD40 plus IL-4 mediated IgE production through alterations of sCD23, sCD54 and IL-6 production. Inflamm Res Official J Europ Histamine Res Soc [et al]. 2005;54(3):113–8. doi:10.1007/s00011-004-1331-8.

42. Schuster GU, Kenyon NJ, Stephensen CB. Vitamin A deficiency decreases and high dietary vitamin A increases disease severity in the mouse model of asthma. J Immunol. 2008;180(3):1834–42.

43. Matheu V, Berggard K, Barrios Y, Barrios Y, Arnau MR, Zubeldia JM, et al. Impact on allergic immune response after treatment with vitamin A. Nutrition Metab. 2009;6:44. doi:10.1186/1743-7075-6-44.

44. Pascual M, Suzuki M, Isidoro-Garcia M, Padron J, Turner T, Lorente F, et al. Epigenetic changes in B lymphocytes associated with house dust mite allergic asthma. Epigenetics. 2011;6(9):1131–7. doi:10.4161/epi.6.9.16061.

45. Perera F, Tang WY, Herbstman J, Tang D, Levin L, Miller R, et al. Relation of DNA methylation of 5'-CpG island of ACSL3 to transplacental exposure to airborne polycyclic aromatic hydrocarbons and childhood asthma. PLoS ONE. 2009;4(2):e4488. doi:10.1371/journal.pone.0004488.

46. Miller RL, Garfinkel R, Horton M, Camann D, Perera FP, Whyatt RM, et al. Polycyclic aromatic hydrocarbons, environmental tobacco smoke, and respiratory symptoms in an inner-city birth cohort. Chest. 2004;126(4):1071–8. doi:10.1378/chest.126.4.1071.

47. Huang TH, Laux DE, Hamlin BC, Tran P, Tran H, Lubahn DB. Identification of DNA methylation markers for human breast carcinomas using the methylation-sensitive restriction fingerprinting technique. Canc Res. 1997;57(6):1030–4.

48. Wu M, Ho SM. PMP24, a gene identified by MSRF, undergoes DNA hypermethylation-associated gene silencing during cancer progression in an LNCaP model. Oncogene. 2004;23(1):250–9. doi:10.1038/sj.onc.1207076.

49. Braun-Fahrlander C, Lauener R. Farming and protective agents against allergy and asthma. Clin Exp Allergy. 2003;33(4):409–11.

50. von Mutius E, Braun-Fahrlander C, Schierl R, Riedler J, Ehlermann S, Maisch S, et al. Exposure to endotoxin or other bacterial components might protect against the development of atopy. Clin Exp Allergy. 2000;30(9):1230–4.

51. Schaub B, Liu J, Hoppler S, Schleich I, Huehn J, Olek S et al. Maternal farm exposure modulates neonatal immune mechanisms through regulatory T cells. J Allergy Clin Immunol. 2009;123(4):774–82 e5. doi:10.1016/j.jaci.2009.01.056.

52. Saetta M, Turato G, Maestrelli P, Mapp CE, Fabbri LM. Cellular and structural bases of chronic obstructive pulmonary disease. Am J Respir Crit Care Med. 2001;163(6):1304–9.

53. Barnes PJ, Shapiro SD, Pauwels RA. Chronic obstructive pulmonary disease: molecular and cellular mechanisms. Europ Respir Journal Official J Europ Soc Clin Respir Physiol. 2003;22(4):672–88.

54. Yang SR, Chida AS, Bauter MR, Shafiq N, Seweryniak K, Maggirwar SB, et al. Cigarette smoke induces proinflammatory cytokine release by activation of NF-kappaB and posttranslational modifications of histone deacetylase in macrophages. Am J Physiol Lung Cell Mol Physiol. 2006;291(1):L46–57. doi:10.1152/ajplung.00241.2005.

55. Moodie FM, Marwick JA, Anderson CS, Szulakowski P, Biswas SK, Bauter MR, et al. Oxidative stress and cigarette smoke alter chromatin remodeling but differentially regulate NF-kappaB activation and proinflammatory cytokine release in alveolar epithelial cells. FASEB J. 2004;18(15):1897–9. doi:10.1096/fj.04-1506fje.

56. Marwick JA, Kirkham PA, Stevenson CS, Danahay H, Giddings J, Butler K, et al. Cigarette smoke alters chromatin remodeling and induces proinflammatory genes in rat lungs. Am J Respir Cell Mol Biol. 2004;31(6):633–42. doi:10.1165/rcmb.2004-0006OC.

57. Lovik M, Hogseth AK, Gaarder PI, Hagemann R, Eide I. Diesel exhaust particles and carbon black have adjuvant activity on the local lymph node response and systemic IgE production to ovalbumin. Toxicology. 1997;121(2):165–78.

58. Moerloose KB, Robays LJ, Maes T, Brusselle GG, Tournoy KG, Joos GF. Cigarette smoke exposure facilitates allergic sensitization in mice. Respir Res. 2006;7:49. doi:10.1186/1465-9921-7-49.

59. Digel W, Lubbert M. DNA methylation disturbances as novel therapeutic target in lung cancer: preclinical and clinical results. Crit Rev Oncol Hematol. 2005;55(1):1–11. doi:10.1016/j.critrevonc.2005.02.002.

60. Ober C, Thompson EE. Rethinking genetic models of asthma: the role of environmental modifiers. Curr Opin Immunol. 2005;17(6):670–8. doi:10.1016/j.coi.2005.09.009.

61. Li YF, Langholz B, Salam MT, Gilliland FD. Maternal and grandmaternal smoking patterns are associated with early childhood asthma. Chest. 2005;127(4):1232–41. doi:10.1378/chest.127.4.1232.

62. Hirahara K, Vahedi G, Ghoreschi K, Yang XP, Nakayamada S, Kanno Y, et al. Helper T-cell differentiation and plasticity: insights from epigenetics. Immunology. 2011;134(3):235–45. doi:10.1111/j.1365-2567.2011.03483.x.

63. Janson PC, Winerdal ME, Winqvist O. At the crossroads of T helper lineage commitment-Epigenetics points the way. Biochim Biophys Acta. 2009;1790(9):906–19. doi:10.1016/j.bbagen.2008.12.003.

64. Janson PC, Winqvist O. Epigenetics–the key to understand immune responses in health and disease. Am J Reprod Immunol. 2011;66(Suppl 1):72–4. doi:10.1111/j.1600-0897.2011.01050.x.

65. van Panhuys N, Le Gros G, McConnell MJ. Epigenetic regulation of Th2 cytokine expression in atopic diseases. Tissue Antigens. 2008;72(2):91–7. doi:10.1111/j.1399-0039.2008.01068.x.

66. Liang SC, Tan XY, Luxenberg DP, Karim R, Dunussi-Joannopoulos K, Collins M, et al. Interleukin (IL)-22 and IL-17 are coexpressed by Th17 cells and cooperatively enhance expression of antimicrobial peptides. J Exp Med. 2006;203(10):2271–9. doi:10.1084/jem.20061308.

67. Stephens GL, Shevach EM. Foxp3 + regulatory T cells: selfishness under scrutiny. Immunity. 2007;27(3):417–9. doi:10.1016/j.immuni.2007.08.008.

68. Shevach EM. Mechanisms of foxp3 + T regulatory cell-mediated suppression. Immunity. 2009;30(5):636–45. doi:10.1016/j.immuni.2009.04.010.

69. Sanders VM. Epigenetic regulation of Th1 and Th2 cell development. Brain Behav Immun. 2006;20(4):317–24. doi:10.1016/j.bbi.2005.08.005.

70. Zhu J, Min B, Hu-Li J, Watson CJ, Grinberg A, Wang Q, et al. Conditional deletion of Gata3 shows its essential function in T(H)1-T(H)2 responses. Nat Immunol. 2004;5(11):1157–65. doi:10.1038/ni1128.

71. Yamashita M, Ukai-Tadenuma M, Kimura M, Omori M, Inami M, Taniguchi M, et al. Identification of a conserved GATA3 response element upstream proximal from the interleukin-13 gene locus. J Biological chem. 2002;277(44):42399–408. doi:10.1074/jbc.M205876200.

72. Loots GG, Locksley RM, Blankespoor CM, Wang ZE, Miller W, Rubin EM et al. Identification of a coordinate regulator of interleukins 4, 13, and 5 by cross-species sequence comparisons. Science (New York). 2000;288(5463):136–40.

73. Solymar DC, Agarwal S, Bassing CH, Alt FW, Rao A. A 3' enhancer in the IL-4 gene regulates cytokine production by Th2 cells and mast cells. Immunity. 2002;17(1):41–50.

74. Ansel KM, Greenwald RJ, Agarwal S, Bassing CH, Monticelli S, Interlandi J, et al. Deletion of a conserved Il4 silencer impairs T helper type 1-mediated immunity. Nat Immunol. 2004;5(12):1251–9. doi:10.1038/ni1135.

75. Lee GR, Spilianakis CG, Flavell RA. Hypersensitive site 7 of the TH2 locus control region is essential for expressing TH2 cytokine genes and for long-range intrachromosomal interactions. Nat Immunol. 2005;6(1):42–8. doi:10.1038/ni1148.

76. Seki N, Miyazaki M, Suzuki W, Hayashi K, Arima K, Myburgh E, et al. IL-4-induced GATA-3 expression is a time-restricted instruction switch for Th2 cell differentiation. J Immunol. 2004;172(10):6158–66.

77. Lee HJ, Takemoto N, Kurata H, Kamogawa Y, Miyatake S, O'Garra A, et al. GATA-3 induces T helper cell type 2 (Th2) cytokine expression and chromatin remodeling in committed Th1 cells. J Exp Med. 2000;192(1):105–15.

78. Yamashita M, Hirahara K, Shinnakasu R, Hosokawa H, Norikane S, Kimura MY, et al. Crucial role of MLL for the maintenance of memory T helper type 2 cell responses. Immunity. 2006;24(5):611–22. doi:10.1016/j.immuni.2006.03.017.

79. Endo Y, Iwamura C, Kuwahara M, Suzuki A, Sugaya K, Tumes DJ, et al. Eomesodermin controls interleukin-5 production in memory T helper 2 cells through inhibition of activity of the transcription factor GATA3. Immunity. 2011;35(5):733–45. doi:10.1016/j.immuni.2011.08.017.

80. Koyanagi M, Baguet A, Martens J, Margueron R, Jenuwein T, Bix M. EZH2 and histone 3 trimethyl lysine 27 associated with Il4 and Il13 gene silencing in Th1 cells. J Biological Chem. 2005;280(36):31470–7. doi:10.1074/jbc.M504766200.

81. Bird JJ, Brown DR, Mullen AC, Moskowitz NH, Mahowald MA, Sider JR, et al. Helper T cell differentiation is controlled by the cell cycle. Immunity. 1998;9(2):229–37.

Chapter 5
Conclusions and Future Perspectives

During the past few years, researchers have witnessed a technical revolution that has dramatically changed the way of doing research in molecular biology. The gene-by-gene studies have been widely replaced by whole genome studies, and both basic and translational areas of research are increasingly getting interested in the role of epigenetics. As a result, epic efforts to develop integrative analysis of biological information keep trying to shed light on the functional and molecular interactions among genome, epigenome, environment, and disease. This is becoming especially relevant for complex diseases as asthma or atopy, which need to be understood as multifactorial diseases.

The epigenome of a cell is the result of specific events that will ultimately determine the regions that need to be expressed, depending on the temporal and environmental circumstances. Virtually, any alteration on the regulation of such tight program could potentially lead to non-physiological states or disease. To better understand the non-genetic variations within a physiological state and then define what is far apart from that condition are essential goals in the epigenetic research field, which also pursues the identification of environmental conditions or aggressions that might be responsible for epigenetic remodeling.

In parallel, the quick rise in the prevalence of allergy, and allergy-related diseases, has lead to postulate theories speculating that environmental factors could influence the individual's predisposition to develop the disease. These theories are based on the observation of the rapid and deep changes that the population has suffered in their lifestyle. Epigenetics has been proposed to hold the key to understand the intricate connections between external environmental aggressions and disease onset and progression.

This promising scenario still holds many challenges and important questions unsolved. Dissecting the different environmental conditions that could modulate individual predisposition to suffer asthma or other allergic diseases is today an open and intriguing area of research. In this sense, to what extent the pathophysiology of allergic sensitization and the severity of the response are mediated by epigenetic mechanisms, or whether these discoveries will lead to novel epigenetic-based anti-allergic treatments are also extremely important questions to be solved. If there are reversible modifications in the epigenome of allergic patients, it would be interesting to know when and how these modifications occurred, if they are

M. Pascual and S. Roa, *Epigenetic Approaches to Allergy Research,* 63
SpringerBriefs in Genetics, DOI: 10.1007/978-1-4614-6366-5_5,
© The Author(s) 2013

inherited or are acquired during the life time, and if we could develop specific targeted therapies to revert or attenuate the malignant epigenetic mark.

Initiatives such as the recently published ENCODE project ([1] and http://www.nature.com/encode/#/threads) to map DNA, protein, RNA, and regulatory elements with the potential to have a functional impact in gene function, can now be explored from any computer in the world. Such public access to epigenetic information is expected to be highly useful to disease research. Researchers can make their own epigenetic maps of cells extracted from patients and healthy controls, and then compare that information to the online available reference maps.

Indeed, current evidences support our optimism that further studies on the epigenetic mechanisms involved in allergy will provide clinicians a set of new tools and biomarkers for the better diagnosis and prognosis of the disease, but most importantly for setting the basis to advance in the prevention and treatment of allergy.

Reference

1. Bernstein BE, Birney E, Dunham I, Green ED, Gunter C, Snyder M. An integrated encyclopedia of DNA elements in the human genome. Nature. 2012;489(7414):57–74. doi:10.1038/nature11247.